The Dazzle of the Digital

The Dazzle of the Digital is written in the context of digital technology's inextricable link with progress and modernity in India, with the COVID pandemic in the backdrop. Digital technology such as smartphones and the internet exemplify the popular ideal of a modernity where the proliferation of data and information seamlessly translates into knowledge and value. The authors attempt to wrestle with this impulsive conflation of the digital with the modern, and argue that the former can sometimes retard progress rather than foster it. They provide examples from various spheres – ranging from public service delivery to private markets – to unpack the pitfalls of a blinkered view on modernity.

The book presents an objective take on the potential of digital technology, written with the hope that it will prompt greater societal reflection on technology as a lever for advancement, at a time when the march of everything digital is inexorable.

Meghna Bal, Fellow, Esya Centre

Vivan Sharan, Partner, Koan Advisory Group

Routledge Focus on Modern Subjects

Series Editor: **Saurabh Dube**, *Professor-Researcher, Distinguished Category, El Colegio de México, Mexico City*

Routledge Focus on Modern Subjects has a broad yet particular purpose. It explores quotidian claims made on the *modern* – understood as idea and image, practice and procedure – as part of everyday articulations of modernity in South Asia, Africa, and the Middle East. Here, the category-entity of the subject refers not only to social actors who have been active participants in historical processes of modernity, but as equally implying branch of learning and area of study, topic and theme, question and matter, and issue and business. Our effort is to explore such modern subjects in a range of distinct yet overlaying ways.

The titles in the series address earlier understandings of the modern and recent reconsiderations of modernity by focusing on a clutch of common and critical questions. Indeed, our bid is to carefully query aggrandising representations of modernity "as" the West, while prudently tracking the place of such projections in the commonplace unravelling of the modern in Global Souths today.

Other books in this series

For more information about this series, please visit: www.routledge.com/Routledge-Focus-on-Modern-Subjects/book-series/RFOMS

The Dazzle of the Digital
Unbundling India Online

Meghna Bal and Vivan Sharan

Routledge
Taylor & Francis Group

LONDON AND NEW YORK

First published 2023
by Routledge
2 Park Square, Milton Park, Abingdon, Oxon OX14 4RN

and by Routledge
605 Third Avenue, New York, NY 10158

*Routledge is an imprint of the Taylor & Francis Group, an
information business*

© 2023 Meghna Bal and Vivan Sharan

The right of Vivan Sharan and Meghna Bal to be identified
as author of this work has been asserted in accordance with
sections 77 and 78 of the Copyright, Designs and Patents
Act 1988.

British Library Cataloguing-in-Publication Data
A catalogue record for this book is available from the British
Library

ISBN: 978-0-367-34303-3 (hbk)
ISBN: 978-1-032-38778-9 (pbk)
ISBN: 978-0-429-32490-1 (ebk)

DOI: 10.4324/9780429324901

Typeset in Times New Roman
by MPS Limited, Dehradun

To Rajiv, Raka, Madhav, Tripurari, and Sujata

Contents

Figures

Tables

Series Editor's Foreword

Saurabh Dube

Routledge Focus on Modern Subjects has a broad yet particular purpose. It seeks to explore quotidian claims made on the *modern* – understood as idea and image, practice and procedure – as part of everyday articulations of modernity in South Asia, the Middle East and Africa. Here, the category-entity of the *subject* also has wide purchase. It refers not only to social actors who have been active participants in historical processes of modernity, but equally implies branch of learning and area of study, topic and theme, question and matter, and issue and business. The series attempts to address such modern subjects in a range of distinct yet overlaying ways.

Questions of modernity have always been bound to issues of being/ becoming modern. These themes have been discussed in various ways for long now.[1] For convenience, we might distinguish between two broads, opposed tendencies. On the one hand, over the past few centuries, it is the West/Europe that has been seen as the locus and the habitus of the modern and modernity. Such a West is imaginary yet tangible, principally envisioned in the image of the North Atlantic world. And it is from these arenas that modernity and the modern appear as spreading outwards to transform other, distant and marginal, peoples in the mould and the wake of the West. On the other hand, such propositions have been contested by rival claims, including especially from within Romanticist and anti-modernist dispositions. Here, if the modern and modernity have been often understood as intimating the fundamental fall of humanity, everywhere, so too have the aggrandisements of an analytical reason been countered through procedures of a hermeneutic provenance.

Needless to say, these contending tendencies have for long each found imaginative articulations, and I provide indicative examples from our own times. The work of philosophers such as Jürgen Habermas and

Charles Taylor and historians such as Reinhart Koselleck and Hans Ulrich Gumbrecht have opened up the exact terms, textures, and transformations of modernity and the modern. At the same time, they have arguably located the constitutive conditions of these phenomenon in Western Europe and Euro-America. In contrast, anti-modernist sensibilities have found innovative elaborations in, say, the "critical traditionalism" of Ashis Nandy in South Asia; and the querying of Eurocentric thought has been intriguingly expressed by the scholars of the "coloniality of knowledge" and "decoloniality of power" in Latin America. These powerful positions variously rest on assumptions of innocence before and outside Europe and the West, modernity and the modern.

Engaging with yet going beyond such prior emphases, recent work on modernity has charted new directions, departures that have served to foreground questions of modernity in academic agendas and on intellectual horizons, more broadly. I indicate four critical trends. First and foremost, there have been works focusing on different expressions of the modern and distinct articulations of modernity as historically grounded and/or culturally expressed, articulations that query *a priori* projections and sociological formalisms underpinning the category-entity. Second, there are the studies that have diversely explored issues of "early" and "colonial" and "multiple" and "alternative" modernity/modernities. Third, we find imaginative ethnographic, historical and theoretical explorations of modernity's conceptual cognates such as globalisation, capitalism, and cosmopolitanism as well as of attendant issues of state, nation, and democracy. Fourth and finally, there have been varied explorations of the enchantments of modernity and of the magic of the modern, understood not as analytical errors but as formative of social worlds. These studies have ranged from the elaborations of the fetish of the state, the sacred character of modern sovereignty, the uncanny of capitalism, and the routine enticements of modernity through to the secular magic of representational practices such as entertainment shows, cinema, and advertising.

Routledge Focus on Modern Subjects engages and exceeds, takes forward and departs from such concerns in its own manner. To start off, its titles address the queries and concepts entailed in earlier explorations of the modern and recent reconsiderations of modernity by focusing on a clutch of common and critical questions. These issues turn on the everyday elaborations of the modern, the quotidian configurations of modernity, on the Indian sub-continent. Next, rather than simply asserting the empirical plurality of modernity and the modern, the series approaches the routine, even banal, expressions of the modern as registering contingency, contradiction, and contention as lying at the core of modernity. Further, it only follows that our bid is not to indolently exorcise aggrandising representations of modernity *as* the West, but to prudently track instead the play of such projections in the commonplace unravelling of the modern in India today. Finally, such procedures not only recast broad questions – for instance of cosmopolitanism and globalisation, state and citizenship, Eurocentrism and Nativism, aesthetics and authority – by approaching them through routine renderings of the modern in contemporary South Asia. They also stay with the dense, exact expressions of modernity yet all the while attending to their larger, critical implications, prudently thinking *both* down to the ground.

In keeping with the spirit of the series, all its titles stand informed by specific renderings – as well as focused rethinking – of key categories and processes. Two exact instances. In different ways, concepts and processes of power and politics alongside those of community and identity variously run through the *Routledge Focus on Modern Subjects*. Here, neither power nor politics are rendered as signifying solely institutional relations of authority centreing on the state and its subjects. Rather, the bid is to articulate these as equally embodying diffuse domains and intimate arrangements of authority and desire, including their seductions and subversions. Actually, as parts of such force-fields, state and government, their policy and programme, might now assume twinned dimensions in understandings of modern subjects. Here can be found densely embodied disciplinary techniques toward forming and transforming subjects-citizens, where such protocols and their reworking by citizens-subjects no less register the shaping of authority by anxiety, uncertainty, and alterity, of the structuring of command by deferral, difference, and displacement.

At the same time, the series approaches community and identity as modern processes of meaning and authority, located at core of nation and globalisation. This is to say that instead of approaching identity and community as already given entities that are principally antithetical to modernity, this cluster explores communities and identities as wide-ranging processes of formations of subjects, expressing collective groupings and particular personhoods. Defined within social relationships of production and reproduction, appropriation and approbation, and power and difference, emergent identities, cultural communities, and their mutations appear now as essential elements in the quotidian constitution, expressions, and transformations of modern subjects.

The Work

The Dazzle of the Digital (henceforth *DoD*) exemplifies our aim to widen the address of modernity and its subjects. On the one hand, this is a book written by fiercely independent development professionals of a younger generation, wedded to practice and thought. Meghna Bal and Vivan Sharan are themselves modern subjects of a specific stripe, graciously contributing to a series chiefly – though not exclusively – populated by denizens of academe. They write with utter sincerity and compelling commitment. Unsurprisingly, the work rubs off the sheen of the dazzle of the digital, its knee-jerk association with modernity and progress. The authors' reveal that such routine rehearsals overlook the structural constraints and messy realities that ought to be addressed by robust, inclusive formations of "development". On the other hand, to a reader such myself – and other like-minded ones – the studies raises a series of key questions. These especially concern the ways in which at the authors' hands the notions of the modern and modernity, technology and progress remain unmarked, even *a priori*, categories-entities. Together, it is the braiding of these twin attributes that signals the fit between *DoD* and the series, especially the wider mandate of *Routledge Focus on Modern Subjects* to endorse and articulate critical heterogeneity, to think and work outside the box.

The principal lineaments of the study concern the projected (yet dashed) promises and the overlooked (yet acute) pitfalls of digital procedures. Here, the precise possibilities of the digital appear betrayed by the uncritical celebration and unthinking adoption of digitalisation, its chants and cants, in the work of governance and administration. All of this actually imbues digital technologies with discrete problems in the business of politics and state. Indeed, policies of digitalisation often entrench and exacerbate structural inequities and commonplace inequalities – of class, caste, and gender as well as of the urban, the rural, and the vulnerable. In such mutual begetting of promise and pitfall, at stake are protocols of technological "evangelism", "substantialist" sensibilities, digital "solutionism", and "remedial" digitalisation. Together, in front are the terms and textures of an instrumental rationality, signalling a fervour and craving for the digital – as fetish and fantasy.

With becoming modesty, Bal and Sharan see their own "contribution" as placing wider critiques of an "almost automatic affinity towards technology and digitalisation" within contexts of "mundane digital policy-related case studies from modern India … [that] straddle various sectors from agriculture to health".[2] At the

xiv *Series Editor's Foreword*

same time, a little later they add that they "also hope to partially offset a general slowdown in the intellectual processes and outputs that have prompted a more holistic and nuanced assessment of modernity in the context of a heterogenous Global South". It is in the gap between what the author's see as their specific contributions to prior critiques of modernity and their broader claims about remedying widespread lacks in current discussions of the phenomena that I now wish to raise a few questions.

But before I proceed further, two qualifications are in order. On the one hand, in the pages ahead my tone might appear sharp, but the effort is toward critical queries that carry forward the search and substance of *DoD* – yet now by way of an animated conversation. Such a dialogue can be had only after a work is written, and it is for this reason that I did not burden the authors with my comments, including suggestions to look up other literatures, at an earlier stage. I feared distracting them from writing their own book. On the other hand, while Forewords of necessity precede a study, I would like to make an unusual suggestion. For all those who are perusing these pages, might I request them to not approach the comments and questions that follow as somehow framing *DoD*, but to return to them after having read the work proper. (That is, unless the kind readers can bear my provocations twice over, now and later!)

Without putting a fine point on the matter, are modernity and progress – and, indeed, technology and development – already known, readily settled categories-processes, as this book insinuates, implicitly and explicitly? Or, is there more to the picture? Here, the author's speak repeatedly of the "march" to modernity and progress, led by meaningfully implemented technology and development, which are all understood as essentially neutral notions and procedures. What such assumption of neutrality implies is that while these motors of history might be viewed in blinkered ways – for instance, as solely contained in GDP as an invariant marker of progress or as primarily embedded in digitalisation as an innate signifier of development – what matters is their "proportionality" to the contexts in which they are deployed.

The following passage brings home these emphases, dramatically.

> Modernity is defined by a kind of sterility that has worked to slowly but surely move organic life out of the realm of human existence and each technological innovation driving it amplifies this decontamination. If we understand modernity to be an evolution of humanity from a primitive, nature-bound species to an urbanized sophisticate, we can look to technology serving as

the primary means of such transformation. And with each technological wave, the messiness of humanity, of being, of life finds itself obsolete or hindering the march of progress. A living being is unpredictable, messy and inadequate in the face of the machine. Machines march on tirelessly, and are therefore, presented as the best versions of organic beings. Thus, human-driven cars replaced human-guided horse carriages, and if all goes well the human will be booted out of the driving equation completely with autonomous vehicles.

Each iteration of technology presents an improved version of its prior avatar. And each wave of technological evolution brings with it an exponential increase in its abilities to solve our problems autonomously, to enhance our abilities to the point that humans are no longer part of the equation. Digital technology is the latest innovation on which we have pegged such hopes. It is purported to solve a range of India's developmental woes. However, the utility of this technology is directly proportional to the way or the *context in which it is deployed* (p. 27, emphasis added).

Actually, it is exactly such framings of modernity and development, technology and progress that have been sieved through imaginative and critical filters across a range of disciplines and knowledges for some time now. Indeed, far from an "intellectual slowdown" concerning "holistic and nuanced assessment[s] of modernity in the context of a heterogenous Global South", there is vast, energetic, and meaningful proliferation of such issues and subjects in contemporary landscapes of thought and theory, practice and performance, methods and methodologies. My reference is to renderings of human sciences and natural ones, work in aesthetics and philosophy, enquiries into the Anthropocene and climate, and departures in Science and Technology Studies (STS). These have revealed science, progress, modernity, and technology to be far more than essentially non-normative, ideal and ideational entities, intimating instead profoundly ideological and nomothetical, fractured and fissured, contradictory and contended, contrapuntal and dissonant procedures of meaning and power – made and unmade by planetary subjects, human and more-than-human.[3]

Again and again, Bal and Sharan return to modernity as the final frontier of history, unravelled by progress, realised through technology, and horizon of development. At the same time, for the authors, what qualifies such scenarios are the requirements of development to be context-sensitive and inclusive, for technology to address "contradictory everyday truths", embodying the Gandhian ideal of *antyodaya*. For it is such inclusivity and ideals that would make whole modernity and progress, setting technology and development on the correct course. These presumptions are written between and across the lines of *DoD*. They announce tensions that are constitutive of the work, about which more later.

My bid now is to pose before Bal and Sharan critical questions concerning modernity, progress, development, and technology – their apparent neutrality, their *a priori* framings. Might we abjure mechanistic projections of modernity in order to approach instead the phenomena as made up of contradictory, conflictual, contested processes of history and society, dominance and desire over the past five centuries? Have not these procedures turned upon – to take only a handful of examples – reason and race, enlightenment and empire, slave-economies and settler-colonialisms, science and genocide, disenchantment and enchantment, colony and nation, commerce and consumption? Are we not in face of protocols of authority and alterity that have been done and undone – not simply by modern subjects but by subjects of modernity?[4]

In such scenarios, does it not make sense to carefully reconsider progress as a trans-historical imperative? On the one hand, can we better approach progress as a particular crystallisation of ideology and action from the mid-nineteenth century, its roots going back to prior developmental blueprints, including of contended strains of the Enlightenment and contrary impulses of the counter-Enlightenment? On the other hand, is it not essential also to acknowledge progress as structures of sentiment and tissues of expectation in our more recently modern worlds, as the authors and assumptions of *DoD* exemplify and that this Foreword too cannot easily escape?

Does not this cluster of queries put a distinct spin on the terms of technology? Can we continue to cast technological matrices as a *deus ex machina*, albeit now in the avatar of an already enchanted yet always expected secular saviour? Or, are technology and technocracy to be unravelled instead as cultures of conquest – of the natural and the social, the human and the non-human, the globe and the planet – that yet formatively bear many meanings, discrete outcomes, more-than-one verities?

Finally, exactly what and how much is at stake in brushing *development* against the grain of its formidable conceits yet along the grain of its pervasive seductions? Are such simultaneous steps not necessary in order to bring forth the rule of experts, the role of denial, and the ruse of ignorance as constitutive of development and its regimes in the present and the past?[5]

While it might seem that I am casting the net much too wide, actually my emphases can provide a different spin to *DoD*. A few examples. Drawing a homology between urbanisation and digitalisation, Bal and Sharan argue that rather being a panacea for development both these processes can reinforce inequities. Yet I am suggesting that instead of seeing such reinforcement of inequality as an unintended consequence, an unexpected outcome, patterns of urbanisation and digitisation are innately complicit upon inequality and hierarchy. It follows that "closeness" and "polarisation" of the well-to-do and the poor in urban spaces are not opposed attributes, as this book implies, but that the intimacies of urbanisation are themselves deeply polarised, as is brought home by the formidable figure of Balram Halwai in Aravind Adiga's *The White Tiger*.

All this is not unlike the manner in which it is insufficient draw a contrast between the closeness of state and citizen, on the one hand, *and* the perpetuation of opacity and corruption, on the other, as part of digitalisation. Actually, it is possibly more important to track how the dyads of state-citizen and opacity-corruption are interwoven with those of governmental-surveillance and its civil-contestations as acute elements of digitalisation. And so, too, rather than apprehend in an either/or manner an improvement in worker productivity and the emergence of pockets of progress amidst hopelessness and despair under urbanisation and digitalisation, it is salient perhaps to consider the mutual entailments of economic efficiency and growing precarity under current regimes of capital and state, especially as mediated by the digital.

I am not splitting hair. At stake in *DoD* is a style of argument that proceeds by presenting the familiar side of the story – around projections of the digital and technology, digitalisation and urbanisation – and then qualifying, questioning, and critiquing such shrill, insidious, self-perpetuating pronouncements. But such a sharp split between rhetoric and reality overlooks their mutual enmeshments in already contradictory worlds. To take an instance, Bal and Sharan turn to amniocentesis as entailing well-meaning technologies that were waylaid by their use toward sex-determination and female foetcide. But there was much more to the picture. In point of fact, the apparently neutral modalities-technologies of amniocentesis actively disregarded gender-social inequalities as part of their exact rationalities-reasons, such that these quietly normative procedures found a remarkable fit with patrilineal ideology and patriarchal practice – simultaneously of state and subject. A corollary to the tragedy and technics of population control – recall the Emergency years – the earlier example is illustrative of wider verities. To wit, the non-ideological, even anti-ideological, stance of technology as a self-realising ideal and practice infuses its techniques and truths with meaning-legislative reasons, which equally find other articulations in quotidian terrains.

Following the authors, under issue of course are "unintended consequences" of policy and programme, technology and digitalisation. Yet, there is a twist to this tale. For such unintended consequences are not simply about a breach between intention and realisation, the overlooking of contexts and ground-realities in matters of application and implementation. On the one hand, we are in the face of a cultivated if varied ignorance, including distinct denials, of unexpected, unknown, and unconsidered outcomes, which are severally built into the imaginaries and practices under discussion. On the other hand, these unintended consequences are not merely one-off occurrences, announcing a slip-up between a particular policy/technology and its resultant effects/ends. Rather, unintended consequences are better understood as the escape of human history from human intention and the return of the consequences of that escape as forming bases for further historical practice. All of this means, for instance, that rather than suggest a somewhat singular connection between the technological evangelism of Nehru and Modi, it is important instead to attend to the slippages and ruptures, the tensions and contortions in such story-lines, including the shifting space-times of everyday imaginings of technology and progress.[6]

The preceding discursus announces distinct readings of *DoD*, their discrete possibilities. Thus, my own endeavour is to approach and apprehend the work *along the grain* of its detailed textures involving agriculture, trade, and health yet *against the grain* of its more preening, prognostic, and palliative analyses. Consider the strengths of the study in laying bare not only the limits of e-governance and digital interventions "targeted at solving two long-standing development concerns [of] agricultural extension and land records", but as these protocols leading up to enhancements and exacerbations of centralised, top-down governance. In front variously are attributes of digital solutionism, technological determinism, and "dashboard-led-development". Behind splashy slogans such as "data is the new oil", these inform the obscuring, indeed exorcism, of heterogenous and unequal demands of land, labour, and capital, which seemingly disappear through the magic of the digital. An innate attribute of governmental modalities, the procedures and processes under issue also bulldoze their way through pandemics and precarities perpetuating grief and misery among the vulnerable in our times of COVID.

To accomplish all this without pulling punches is impressive achievement. Here is a work that invites us to be ever cognisant of contradictions of worlds, which make urgent demands on us. Its authors equally impel us to stay with and think through such wider contrariness. To engage such spirit, substance, and sensibility of the book is to discover Bal and Sharan as offering imaginative ethnographic accounts of state and governance, ahead of endless alchemies of digital technologies and developmental fantasies.

At the end, *DoD* is at once an optic and a register. If we do not make that ready separation between subject and object, the expert and the world, the authors of this book – and the writer of this Foreword – do not emerge as somehow autonomous analysts. Rather, we are all subjects of the digital, the technological, the developmental and of course much else. To be found amidst ourselves are subjects and technologies entirely enmeshed in the contentions and contrariety of modernity, the blinkers and seductions of progress, already always in quotidian keys.

Ready resolutions are easy to find but difficult to sustain in worlds that have been, terrains that are, posterities that might be. This is one among the many lessons to be learnt from Meghna Bal, Vivan Sharan, and their *Dazzle of the Digital*.

Notes

1 The discussion in this Foreword of different understandings of modernity (and the modern) draws upon a wide range of scholarship. Instead of cluttering the short piece with numerous references, indicated here are a few of my works that have addressed these themes – in dialogue with relevant literatures – and that back the claims made ahead. Needless to say, prior arguments and emphases are being cryptically condensed and radically rearranged for the present purposes. Saurabh Dube, *Subjects of Modernity: Time/Space, Disciplines, Margins* (Manchester: Manchester University Press, 2017); Saurabh Dube, *Stitches on Time: Colonial Textures and Postcolonial Tangles* (Durham and London: Duke University Press, 2004); and Saurabh Dube, *Disciplines of Modernity: Archives, Histories, Anthropologies* (London and New Delhi: Routledge, forthcoming 2022). Consider also, Saurabh Dube, *After Conversion: Cultural Histories of Modern India* (New Delhi: Yoda Press, 2010), Saurabh Dube (ed.), *Enchantments of Modernity: Empire, Nation, Globalization* (New Delhi and London: Routledge, 2009, 2010); and Saurabh Dube (ed.), *Handbook of Modernity in South Asia: Modern Makeovers* (New Delhi: Oxford University Press, 2011).

2 In keeping with the style and sensibility of the Forewords to the titles in this series, I do not offer a detailed summary of the arguments of *DoD*, which the work does not require in any case. At the same time, the wider emphases and orientations of the book are discussed in the pages ahead.

3 A handful of these shifts and sensibilities were discussed in the first part of
  the Foreword. Limitations of space prevent me from any further explication
  or even indicative references, which would easily run into a few dozen at the
  least. Here, I am happy to indicate that several of the books in this series –
  published, forthcoming, and contracted – broach such issues in their own
  way. These include especially titles by Pankaj Sekhsaria, Gilberto Conde,
  Projit Bihari Mukerjee, Navtej Johar, Saladdin Ahmed, and my own intervention.
4 I discuss such questions in, for instance, Dube, *Stitches on Time*; Dube,
  Subjects of Modernity; and Dube, *Disciplines of Modernity*.
5 Once more, the preceding discussion is based upon a wide variety of critical
  considerations. Some of these issues are to be addressed in Saurabh Dube,
  Zine Magubane, and Prathama Banerjee, *Decolonize: Three Enquiries in
  Discipline and Difference*. Manuscript of book in progress.
6 The discussion in Chapter 2 of the development of technology in independent India is intriguing. As I understand, the authorial effort is to
  unravel technological evangelism and its unintended consequences from
  Nehru through to Modi. Yet the developments under Nehru are presented
  as a sort of "heroic history", even though one of the two principal sources
  for the storyline, an essay by the historian David Arnold, casts a sceptical,
  indeed ironic, eye on the subject. Two points follow. On the one hand, recall
  my point earlier regarding progress as a structure of sentiment, a tissue of
  expectation, a texture of affect: it is such sense and sensibility of progress –
  fortified by pedagogy, institutional and everyday – that arguably orches-
  trates the expansive account, its intrepid attributes, of technology under
  Nehru in *DoD*, despite the authors' own assumption and analytic. On the
  other hand, reading the manuscript, I confess to having wrestled with the
  conundrum of whether or not to invite Bal and Sharan to rethink this
  narrative, especially revisiting Arnold's essay as well as reading other critical
  writings on the subject. In the end, I refrained for the reason that tensions
  and contentions in argument are more productive to ponder than sanitised
  stories.

Acknowledgements

We would like to thank Saurabh Dube, the Series Editor, for giving us the opportunity to write this book and guiding us so patiently through the proposal process. We are grateful to our dogs Tara, Luna, Harper, Trotsky, Chutki and Husky for providing us with both welcome and not-so-welcome distractions during the writing process. We couldn't have done this without you. And yet, we also feel that we somehow managed to get our writing done despite you, particularly Harper. Finally, we would like to thank Vibodh Parthasarathi, Ananth Padmanabhan, Megha Bahree, Deepak Maheshwari and Aafreen Ayub for their constructive criticism and support.

1 Introduction – The Dazzle of the Digital

We write this book in the midst of the COVID pandemic, at a time when the dazzle of everything that is "digital" is at its peak. The notion of "digital" represents much more than bits and bytes of computer data today. It exemplifies a popular ideal of a modernity where the proliferation of data and information seamlessly translates into knowledge and value, which lift all boats. However, we simultaneously attempt to wrestle with this impulsive conflation of the digital with the modern, and argue that the former can sometimes retard modernity rather than foster it. It is important for Indian society and polity to understand their relationship better, to avoid a blinkered or idealistic view that has led to unintended consequences in the past. Else, we may perpetuate socio-economic inequity, and consequently find it harder to break structural gridlocks to inclusive and sustainable development.

India and the world are celebrating digitalisation for the fact that it allowed societies and states to communicate, transact, work, and govern, despite physical restrictions to contain the spread of the pandemic. The Secretary General of the United Nations (UN) has consequently termed digital connectivity "indispensable, both to overcome the pandemic, and for a sustainable and inclusive recovery" (Guterres 2020). Similarly, the UN's Sustainable Development Goals highlight that the

> spread of information and communications technology and global interconnectedness has great potential to accelerate human progress, to bridge the digital divide and to develop knowledge societies, as does scientific and technological innovation across areas as diverse as medicine and energy. (UN General Assembly 2015)

From being a good to have, digital has become a must-have. It is true that if the pandemic struck in the nineteen nineties, when digitalisation

DOI: 10.4324/9780429324901-1

and the internet were at their infancy, it would have crippled the global economy and stalled progress for much longer than today.

Luckily, over the last decade or so, a majority of Indian citizens are now connected to broadband. This is no mean feat, achieved through privatisation of telecommunications, explosion of affordable devices and a hypercompetitive digital economy. Consequently, nearly 12 percent of the people using the internet globally belong to India, which represents the second largest user base in the world after China and ahead of the US (Grover 2019). The combined growth of digital infrastructure, devices and applications over this period, represents unbounded opportunity.

Digitalisation has brought communities closer, much like urbanisation has. Urban areas account for about a third of India's population using the Census's stringent definitional criteria, however, satellite images show that this share is much larger – perhaps closer to two-thirds of the country (Sreevatsan 2017). Both phenomena have helped to connect state and society, enmeshed the personal and the professional, and liberated many millions from the tyranny of geography. But, the likeness between digitalisation and urbanisation does not end here.

Urban areas represent a larger gap between the haves and have-nots than rural areas. For instance, urban wage inequality is much higher than rural wage inequality (Alvaredo et al. 2018). The bottom 40 percent of population groups in urban areas saw the slowest increase in per capita consumption expenditure between 1994 and 2012, whereas the top 20 percent saw the fastest (Himanshu 2018). That is, the urban poor are worse off than the rural poor in terms of relative inequity in progress. Urban inequality is multidimensional and complex, and results from various factors such as: "education inequality, transportation inequity, spatial mismatch, environmental injustice, the digital divide, food deserts, and unequal access to government services" (Nijman and Wei 2020). The prevalence of these diverse factors makes urban inequality a hard challenge to overcome. Similar divergences are likely to manifest between digital haves and have-nots, if associated transformations and transitions aren't managed well. The UN Secretary-General also considered this dimension in stating that "unless we address digital instability and inequality, they will continue to exacerbate physical instability and inequality" (Guterres 2020). In fact, experts contend that the emergence of the digital economy, where incomes tend to be higher than in traditional manufacturing or services, has led to a "bifurcated workforce and a polarised social structure" (Nijman and Wei 2020).

That is, the phenomena of urbanisation and digitalisation can re-inforce inequality, even though they are simultaneously well-accepted markers of progress. We try to understand the implications of this contradiction between the progress and regression in the context of a fast-digitalising India, and suggest means to identify and avoid the pitfalls of a blinkered view on modernity.

India must heed the warning signs that portend the fact that digital technologies will exacerbate entrenched developmental inequities. We focus on a few areas that best represent this problem. In the next chapter, we show that the country's history is replete with ex-amples of unintended consequences of technological evangelism, that foreshadow similar outcomes in the digital age. In the third, we focus on how reliance on e-governance can heighten substantialist tendencies in government and therefore add complexity to already messy development challenges. In the fourth chapter, we highlight how the exuberance around the digital economy, has led to a form of the "Dutch Disease", wherein digitalisation is seen as a palliative for underdevelopment of traditional factors of production like land, labour and capital. And finally, in the fifth chapter, we illustrate how a fervour for digital, led to the exclusion of the millions from much-needed healthcare interventions and social safety nets in the midst of the pandemic. We conclude that the digital divide is real, and is a structural barrier to the Gandhian ideal of Antyodaya.

None of the themes we cover in this book are meant to spark the inference that India should not digitalise (or urbanise). Far from it. Like urbanisation, digitalisation is inevitable. It also demands good policy design and management to bear fruit. And these elements re-quire governments, industry and civil society to begin to account for the paradoxical facets of digitalisation. For instance, like urbanisation, unmanaged digitalisation brings people closer but also polarizes them. It seemingly reduces the separation between the State and citizens but does this while preserving and sometimes even perpetuating opacity and corruption. Like urbanisation, it can improve worker productivity and economic competitiveness, or create islands of progress amidst seas of hopelessness and despair.

We therefore explore the central paradox of digitalisation and di-gital technology – that while they are full of promise in the developing country context – they can also make existing problems tougher. Their dazzle can confuse the means and the ends of developmental interventions and market growth. Their automatic association with everything modern can hide contradictory everyday truths. There is a need to recognise this reality in the Indian context, in order to solve it.

Four problematic characterisations

Digitalisation, or the process of converting something into digital form, is no panacea. But the temptation to see it as such is immense. Such an impulse is particularly acute in governance and politics, because of digitalisation's universal appeal as a vehicle for transformation of societies and markets. That digitalisation serves as a cure-all for development is often reinforced by international organisations like the World Bank, which contends that "Governments can meet their constituents' expectations by investing in a comprehensive public-sector digital transformation" (Melhem 2019).

It is also common for digital technology to be seen as a silver bullet for efficiency in the private sector – particularly in a world where customer reach alone can generate corporate value. According to the PwC's 24th Annual Global CEO Survey, "half of CEOs plan increases of 10 percent or more in their long-term investment in digital transformation" in the post-COVID scenario (Boswell et al. 2021). Therefore, we see "digital transformation" from a broad lens – one that extends much beyond the integration of digital technology into either governmental or business processes exclusively. We refer to it as a term that represents the process of integration of digital technology to change and improve the functioning of organisations, cultures, and entire ecosystems, that necessarily involves both the public and the private.

However, we contend that asymmetric focus on the "digital" in digital transformation, can make hard problems harder, whether in government or the private sector. This is because there are at least four popular characterisations of digital that are problematic. Each makes organisations and governments focus on technology as an end in itself.

The first such characterisation, which digital technology shares with the larger world of technology, is an uncontested association with modernity. This impulse to conflate the two is problematic because it can hide underlying causes of real problems that digital technology may aim to solve. From Nehru to Modi, India's Prime Ministers put personal stakes in flagship programmes on technology. They therefore fanned a "technological determinism" a term coined by American economist and social scientist Thorstein Veblen, and one that centres on the proposition that technology determines the nature of societies. An affinity for technology has made for great leadership branding – since most political figures associated with technology in the developing world are considered reformist. Therefore, in the second chapter, we argue that technology makes for a pliable political messaging tool. Political messaging imbues technology with innate goodness as a bridge between the rich and the poor, the marginalised and the empowered, the worker and the capitalist, the state and its people.

Of course, technology's goodness is not absolute, and the context in which it is applied as a solution, matters. For instance, we outline how the introduction of amniocentesis technology in India in the mid-seventies, ostensibly to control population growth, eventually led to a skewed sex ratio. The technology was introduced through American technical and grant assistance to the All India Institute of Medical Sciences, on the premise that it could aid diagnosis in cases of inter-sexuality and infertility. However, its application and spread as a tool for sex determination and eventually female infanticide was an unintended consequence. Modern India is replete with accounts of similar unintended consequences that resulted from application of digital as a cure-all for challenges, without appropriate consideration of underlying context.

A second popular characterisation of digital is that it can help achieve governance interventions at scale. However, we contend that there is a simultaneous de-emphasis on the individual, the smallest yet most perhaps a most important unit around which inclusive and equitable development interventions should be located. In this context, we discuss the effects of digitalisation on governance in India. The telecommunications revolution quickly transformed into a digital one, with close to a billion internet users in the country. Naturally, development interventions via e-governance in a country with challenges at such a large scale cannot be homogeneous. However, there are some common fibres that characterise governance efforts in India, particularly in caricaturisation of information-led development into what we dub as "dashboard-led development". Dashboard-led development speaks to a paradigm where the more digitally linked a state is to its constituents, the less connected it is with what is happening on the ground. Much like other data-led development models, dashboard-led development is about managing upwards – a way for departments to showcase "results" without dealing with the intractability of messy policy concerns.

It may be worth noting that dashboard-led development is akin to "digital solutionism". The latter is a term coined by the scholar Evgeny Morozov, who summarily dismisses the "romantic utopia" that a new "problem-solving digital infrastructure" can be uniformly employed in all circumstances, and remain effective. Chief amongst the risks of this approach is the concomitant under-emphasis on designing for the needs of individual information recipients, based on the presumption that the two-way nature of technology is a sufficient tool to build useful knowledge. We find this to be a key miscalculation that is common to e-governance interventions. We analyse the premise that individual interests can be automatically met through the use of digital technology, and that too at previously unprecedented scale in the third chapter.

A third characterisation of digital technology is that it drastically reduces the cost of distribution of goods and services, and overcomes several limitations of traditional markets. The challenge with this conception is that it melds fact and fiction, and therefore breeds confusion. It conflates the tangible benefits of digital technology for trade and commerce, with the notion that these will come naturally, no matter what the policy design and ideological direction. The premise of the characterisation is that digital technology naturally fosters economic competitiveness in two ways. It allows brands to market their offerings directly to target consumers through the internet, rather than through retail intermediaries that are physically bound. This allows producers to capture additional margins and offer superior quality at lower costs. It also allows consumers to better assess product fit and easily ascertain relevant product information. For instance, studies show that a large segment of potential consumers refers to company websites or eCommerce platforms before making purchases ("Digital Transformation B2B E-Commerce 2020–2028" 2020). Additionally, digitalisation allows businesses to interact with each other seamlessly, and consequently has led to the emergence of a wide variety of business-to-business (B2B) solutions. In fact, the B2B eCommerce market was estimated at $1.2 trillion in 2021, in the case of the US economy alone (Walker 2018).

Therefore, we focus our fourth chapter on the dazzle of digital in the context of economic competitiveness. We find that the promise of digital manifests most prominently in the now popular concept that "data is the new oil". However, digital markets effectively put businesses from across the world on a truly level playing field, where geographic limitations of traditional businesses are no longer a constraint. This means that even though digital fosters a new competitiveness through its associated distributional efficiencies described above, it also forces Indian businesses to compete in a new market paradigm, with global counterparts. We highlight the complications this throws up in a country like India where traditional factors of production such as land, labour and capital remain underdeveloped, and therefore markets and businesses remain underequipped to make the most of digitalisation. The rhetoric around data and oil doesn't help address this gap, and perhaps makes it worse, because it leads to an underemphasis on good policy design/vision and an overemphasis on protectionism. Consequently, India's top 10 digital companies are valued at about 10 percent of Chinese equivalents.

A fourth and final purported characteristic of digital technology is its ability to help identify beneficiaries and target developmental interventions at vulnerable sections of society. We challenge this notion because we find that digitalisation of development interventions also runs the risk of ignoring those from within vulnerable sections, who are most at risk. That is, digital tends to aggregate the vulnerable, which masks the challenges of the most ill-protected by social schemes. The characterisation of digital as innately inclusive stems from various reputed quarters. For instance, the UN's Sustainable Development Goals stress the need to enhance the use of digital technology to "promote the empowerment of women" (Goal 5), and therefore position universal access to digital technology in the poorest countries as an objective in itself (Goal 9). In fact, digital is now a mainstay of development literature, as a tool that can help spread the benefits of social progammes more evenly, capture previously unavailable data, and offer an unparalleled transparency and accountability in basic service delivery. As an aside, the use of digital for development represents one of the few areas of consensus between the developed countries that represent the Global North and less developed counterparts in the Global South. This is perhaps because of the presupposition that unlike earlier general-purpose technologies, substantive digital capabilities exist in the South (Joseph 2004).

We analyse this strong use case made out in favour of digital for development in the fifth chapter. But we do so in the specific context of the COVID pandemic, in order to test the efficacy of digital interventions when they were most needed and in the case of those who needed them the most. For instance, we look at the effectiveness of health surveillance through digital applications, as well as social welfare transfers through digital infrastructure. Unfortunately, we find that the use of digital ended up exacerbating existing inequities in access the health and finance, and negatively impacted those at the bottom of the socio-economic pyramid in India. The findings of this chapter in the context of a global emergency were telling, because they illustrate that digital can sometimes work at cross purposes with the Gandhian ideal of Antyodaya – or the targeted upliftment of the weakest sections of society. A modern translation of this ideal can also be understood as a call to arms to design development interventions from the ground up – an exercise that requires much greater effort than to design for aggregate groups and problems in the abstract. We recommend that greater attention be paid to notions like account-ability, state capacity and user trust, all elements that are often taken for granted in the design of digital development.

In fact, the romanticisation of digital in the development context requires urgent rethinking, across the board. For this, India's governing and intellectual elite must re-examine the four conceptual fallacies outlined above. That is, they will need to avoid conflating the digital with the modern; resist the temptation to view e-governance as a cure-all; put greater not lesser emphasis on policy design to leverage digital technology towards economy competitiveness; and above all, focus on those most at risk from being left out of a new digitalised modernity.

We recognise that most of the characterisations and their conceptual rebuttals that we propose are also discussed by eminent thinkers and even world leaders across diverse temporal and developmental contexts. These include works of contemporary scholars such as that of Evgeny Morozov cited earlier, as well as the timeless critiques of techno-economic conceptions of modernity by leaders like Mahatma Gandhi. Our contribution to the existing body of work that questions an almost automatic affinity towards technology and digitalisation is therefore to place such questions in the context of mundane digital policy-related case studies from modern India. These straddle various sectors from agriculture to health. The country aims to provide universal access to broadband internet to all its people, even as it endeavours to provide all citizens access to much more basic and essential necessitates such as drinking water and electricity. It's the only country in the world that seeks to do all of this at the scale of over a billion people since China's political and economic structure allowed for this feat in a more ordered and seemingly logical sequence. In grappling with technology and modernity through this lens, we also hope to partially offset a general slowdown in the intellectual processes and outputs that have prompted a more holistic and nuanced assessment of modernity in the context of a heterogenous Global South.

Economic growth and modernity

Established discourses on economic growth and modernity also foreshadow more contemporary critiques of technology and modernity, and more specifically in our context, digital technology and modernity. The similarities between these discourses and their respective trajectories are uncanny, and deserve to be correlated to the context of this Focus Series on Modern Subjects, albeit briefly.

Economic growth, as measured by Gross Domestic Product (GDP), is a widely accepted measure of progress and modernity. The concept of a GDP was developed by Simon Kuznets, a Russian-born American economist in the 1930s. The idea was to capture all economic production or output in a single measure. And in many ways, this was an elegant indicator, since it would rise in good times, and fall during the bad. Kuznets felt the need for an all-encompassing indicator to understand the full impact of the Great Depression (Kuznets, 1932). The construct became part of the global economic zeitgeist following the Bretton Woods Conference in 1944, where 44 nations charted out the future of the international monetary system. These discussions were dominated by the US and Britain but the outcomes, including the use of GDP as the main measure of a country's economy, were accepted by all. This is in consonance with many of the routine articulations of technology and modernity, which tend to stem from the West but are widely accepted. This includes the purported centrality of digital technology in nations becoming "modern".

The irony in the centrality of the GDP as a measure of progress in the modern age is that even Kuznets warned against its indiscriminate use. Moreover, several American economists, such as Moses Abramovitz also cautioned against the use of GDP as a measure of individual well-being or societal welfare. He highlighted that changes in the rate of growth of economic output were not a good measure of welfare in the long-term (Dickinson 2011). It is easy to understand why. For instance, in most developed countries, consumption is the largest component of GDP. However, it is unclear whether all consumption results in productive growth that can lift all boats. These types of first-principle questions were in sharp focus in the aftermath of the Global Financial Crisis, for a brief while. We can perhaps expect such scepticism, contestations and pushbacks every time there is a catastrophic failure in global economy or even in case of the environment or ecology as a result of misgovernance of global commons.

We see a similar trend in the case of discourses that conflate the digital with the modern. That is, much like economic growth as measured by the GDP has become the de facto indicator of progress and well-being, even though was is not meant to be seen as such, we also see digitalisation employed as an all-encompassing indicator of progress, rather than as a tool that can be used to foster progress. That is, it is seen as an end in itself. This is particularly the case in the aftermath of the COVID pandemic. Digital adoption is now seen as a panacea to economic slowdown and social distress, which makes it tempting to draw another analogy with the role of the GDP post the Great Depression. For instance, the International Monetary Fund (IMF) posits that

> despite all the economic and social devastation it (COVID) has caused, (it) provides an opportunity for African countries to innovate and go digital. African countries will have to rebuild their economies. They should not merely repair them; they should remake them, with digitalization leading the way. (IMF, 2021)

With scant details on what this means for physical systems, or structural and administrative challenges, this is nothing if not digital solutionism.

Moreover, as in the case of the GDP, the prominence of digital or more precisely, digitalisation, as an indicator of progress, is common to both Global North and South. This is despite development literature that may point towards the pitfalls of a blinkered approach or one that is ill-suited to the needs of the Global South. For instance, the Indian-origin economist and Nobel Laureate Amartya Sen spent a large part of his career evangelising measures of "Human Development" as alternatives to the GDP or other similar aggregate indicators like the Gross National Product (GNP). The Human Development Index, which he champions, has three components – indicators of longevity, education and income per capita. These, he says, "broaden substantially the empirical attention that the assessment of development processes receive" because they are "not exclusively focused on economic opulence" (UNDP 1999). Of course, he is right. There is no case to be made that the GDP should be used as a proxy for welfare or inclusive development by itself. That's because GDP growth can coexist with the growth of inequity, much like digitalisation co-exists with digital divides. However, it is human to attempt to simplify our world. And the use of an indicator like GDP, or the level of "digitalisation" of a society, is much easier than to measure the well-being of the most deprived or to quantify the comparative advantages of digital industries across closely integrated global markets.

Therefore, we explore the many contradictions that are at the core of the characterisation of the digital as the modern. We do so through a preponderance on issues related to public policy and technology, so as to put into sharp relief the prevalence of the confusion among the elite who have agency and the power to rob others of it. And this policy orientation across different sectors also allows for a more contemporary and dynamic assessment in a field that is constantly evolving, as digital technology is itself. And finally, we also recognise that greater deliberation on these contradictions may require a more considered and nuanced public debate. An additional reason why we think the themes in this book are relevant is that general public debate on digital technology is skewed towards social media and digital communications globally and in India, despite their negligible relevance to development outcomes.

Rebalancing public debate

The fact that digital technology like social media occupies a large share of mind space is not surprising. This is especially because its very large user base is very invested in social media (India itself has over half a billion such users), despite its association with several negative characteristics that breed mistrust in society. For instance, seven-in-ten Americans use social media (Anderson and Auxier 2021), even though around two-thirds say it has a "mostly negative effect on the way things are going" in terms of the polarisation of society (Anderson and Auxier 2021).

Similarly, the world's largest survey of internet security and tryst, including of Indians, found that "social media companies were second only to cyber criminals when it came to fuelling online distrust" (CIGI-Ipsos Global Survey 2019).

Some of the most adverse impacts of digitalisation are indeed visible in social media applications that enable information sharing. Well-documented phenomena like fake news and disinformation are prominent on such media. In fact, we write this at a time when the outgoing US President Donald Trump was accused of using his social media handles to instigate mobs to storm Capitol Hill. For this purported misuse of social media, he was banned from several such platforms. Ironically, Trump also issued an Executive Order in 2020, that termed social media as modern-day equivalent of the Public Square. That is, he suggested that social media should be compared to physical public spaces where individuals are held to account for any speech that instigates violence or bad behaviour. These are not superficial debates, since technology and the battle for information in democracy deserve our full attention. Moreover, societies must engage with this theme because, for a long time, social media was unassailable. However, this façade has worn off.

Nevertheless, social media is the tip of a very large iceberg of information societies. The rest of the iceberg deserves its fair share of attention because it is also enmeshed in our lives in less visible but equally important ways.

Indian society and its polity must focus on the less glamourous aspects of digital technology that often escape debate. We analysed questions in the lower house of the Indian Parliament, from 2014 to March 2021, to understand the level of attention paid to social media over other pervasive themes linked to digitalisation. We did so through a frequency analysis of key terms used for questions raised in Parliament. Table 1.1 contains our findings. We see that the Parliament's focus on social media saw a sharp uptick in the second term of the Modi Government (2019 onwards). This was coterminous with several adverse events that had impacted the global debate on digitalisation, including the Cambridge Analytica scandal in the West. Simultaneously, more fundamental issues linked to digital inclusion like online payments, as well as deeper aspects of digital such as on software, cloud services and artificial intelligence (reflected in Table 1.1 under the heading of technology) received scant attention.

We write this book in the hope that the brick and mortar of digital, the non-glossy, non-dazzling aspects, yet those that promise to substitute, supplement and supersede the physical, are also debated with equal vigour in the days ahead. We therefore write about digital technology's undergird, the pillars on which governance, public

Table 1.1 Changes in Focus of Lok Sabha on Issues Linked to Digitalisation[1]

Theme	Share in Second Term (percent)	Share in First Term (percent)	Change in Share (percent)
Digital Literacy	2.15	–	–
Social Media	13.88	2.89	10.99
Data Protection	12.68	6.27	6.41
Infrastructure	3.83	0.8	3.03
Internet Restrictions	3.35	0.96	2.39
China	2.39	0.16	2.23
Fake News	2.15	0.96	1.19
Government Programmes	16.75	15.92	0.83
Child Protection	0.96	0.48	0.48
Gaming	0.96	0.64	0.32
Online Payments	1.91	3.7	−1.79
Aadhaar	9.09	12.22	−3.13
Technology	37.56	44.86	−7.3

Source: Authors' Calculations Based on Government Database of Lok Sabha Questions.

service delivery and commerce will develop in the 21st century. We do so not to make the case against digitalisation, or modernity, but for calibrated versions of it. We question the dazzle of digital, and treat its automatic association with progress with sceptical lens, because unless India addresses the propensity of digital solutionism to override structural and systemic concerns, it may do more harm than good. Conversely, if the unintended consequences of digital solutionism are not discussed, a consequent breakdown of public trust in technology may precipitate its own vicious cycle of scepticism and cynicism. That would be an unfortunate outcome, and one that will throttle the pace of technology adoption and exacerbate the digital divide.

Note

1 Number of questions in each theme counted using the number of times keywords related to these themes were mentioned in the title of the question asked to the Ministry of Electronics and Information Technology in India's Lok Sabha. The most recent question asked to the Ministry in the 17th Lok Sabha is dated 24 March 2021.

Bibliography

Anderson, Monica, and Brooke Auxier. "Social Media Use in 2021." Pew Research Center, April 7, 2021. https://www.pewresearch.org/internet/2021/04/07/social-media-use-in-2021/

Alvaredo, Facuno, Lucas Chancel, Thomas Piketty, Emmanuel Saez, and Gabriel Zucman. "World Inequality Report." World Inequality Lab, 2018. https://wir2018.wid.world/files/download/wir2018-summary-english.pdf

Assembly, UN General. "Transforming Our World: The 2030 Agenda for Sustainable Development." United Nations, October 21, 2015. https://sdgs.un.org/2030agenda

Boswell, Libby, Spencer Herbst, and Catherine Moore. "PWC 24th Annual Global CEO Survey: A Leadership Agenda to Take on Tomorrow." PWC, 2021. https://www.pwc.com/gx/en/ceo-survey/2021/reports/pwc-24th-global-ceo-survey.pdf

CIGI-Ipsos Global Survey on Internet Security and Trust' 2019. Centre for International Governance Innovation and United Nations Conference on Trade and Development.

"Digital Transformation B2B E-Commerce 2020–2028." Absolute Market Insights, November 2020. https://www.absolutemarketsinsights.com/reports/Digital-Transformation-B2B-E-commerce-2020–2028-744

Dickinson, Elizabeth. "GDP: A Brief History." Foreign Policy, January 2011. https://foreignpolicy.com/2011/01/03/gdp-a-brief-history/

Duarte, Cristina. "Africa Goes Digital." IMF, 2021. https://www.imf.org/external/pubs/ft/fandd/2021/03/africas-digital-future-after-COVID19-duarte.htm

Feijóo, Claudio, Jose Luis Gómez-Barroso, and Edvins Karnitis. "More than Twenty Years of European Policy for the Development of the Information Society." *Les Politiques Supra-Nationales En Europe* 21–1, no. 2 (2007): 9–24. 10.4000/netcom.2389

Grover, Divya. "India Has World's Second-Largest Internet User Base." *Bloomberg Quint*, June 12, 2019, Online edition. https://www.bloombergquint.com/technology/india-has-worlds-second-largest-internet-user-base

Guterres, Antonio. Secretary-General's remarks to the Virtual High-Level Event on the State of the Digital World and Implementation of the Roadmap for Digital Cooperation, New York, June 11, 2020, https://www.un.org/sg/en/content/sg/statement/2020-06-11/secretary-generals-remarks-the-virtual-high-level-event-the-state-of-the-digital-world-and-implementation-of-the-roadmap-for-digital-cooperation-delivered

Himanshu. "India Inequality Report 2018: Widening Gaps." India: Oxfam India, 2018. https://www.oxfamindia.org/sites/default/files/WideningGaps_IndiaInequalityReport2018.pdf

"Human Development Report." United Nations Development Programme (UNDP), 1999. http://www.hdr.undp.org/sites/default/files/reports/260/hdr_1999_en_nostats.pdf

Joseph, K.J. "Transforming Digital Divide into Digital Dividend: The Role of South-South Cooperation in ICTs." Research and Information System for the Non-Aligned and Other Developing Countries, 2004. https://www.ris.org.in/transforming-digital-divide-digital-dividend-role-south-south-cooperation-icts

Kuznets, Simon. "National Income, 1929 – 1932." National Bureau of Economic Research, Bulletin 49, 1934. https://www.nber.org/system/files/chapters/c2258/c2258.pdf

Melhem, Samia. "Assessing Country Progress Towards Digitization." World Bank Group, February 2019. https://pubdocs.worldbank.org/en/658001549914908574/TransportGlobalPractice-Note-February-web.pdf?deliveryName=DM10670

Morozov, Evgeny. "To Save Everything, Click Here: The Folly of Technological Solutionism." PublicAffairs, 2013.

Nijman, Jan, and Yehua Dennis Wei. "Urban Inequalities in the 21st Century Economy." Elsevier, Applied Geography, Volume 117, April 2020. https://www.sciencedirect.com/science/article/pii/S0143622820301910?via%3Dihub

Sreevatsan, Ajai. "How Much of India Is Actually Urban?" *Mint*, September 16, 2017. https://www.livemint.com/Politics/4UjtdRPRikhpo8vAE0V4hK/How-much-of-India-is-actually-urban.html

Walker, Brian. "On the Verge." Accenture, 2018. https://www.accenture.com/_acnmedia/pdf-78/accenture-verge-b2b-digital-commerce.pdf

2 A Brief History of the Politics of Technology in India

In October 2015, an Indian bureaucrat gave a presentation on "Digital India", an initiative to enable the electronic delivery of government services to all citizens, to a group of World Bank officials in Washington D.C. (Yadunath 2016). He detailed the vision for Digital India as a tripartite stack. Digital infrastructure was envisioned as a utility for every citizen and sat at the base layer of this stack (Sharma 2015). This infrastructural foundation encompassed the provision of ubiquitous high-speed internet, a digital identity programme, and universal access to mobile phones and bank accounts to enable greater financial inclusion and empowerment. Such baseline infrastructure could then set the stage for the second layer of the stack, namely e-governance and public services on demand. That is, the seamless integration of different government departments through technology and the instant delivery of public services to even the most remotely located citizen (Sharma 2015). Finally, the third part of the stack encompass digital empowerment of citizens through literacy and awareness initiatives, widespread availability of the digital resources, coupled with a digital landscape that is empathetic towards the country's linguistic heterogeneity, and a participative democracy enabled through digital platforms that offer multiple services (Sharma 2015). What was described to the World Bank sounded logical and seemingly solved structural challenges linked to India's development journey.

The "stack" of digital solutions was a stack of promises. It represented promises of cheaper, faster and easier governance, to replace an overstretched brick-n-mortar framework. The stack was therefore a digital edifice that could potentially meet several political and national objectives. It doesn't come as a surprise that most major national parties reflect on digital technology in their political manifestoes. In fact, the two major national parties that are usually guaranteed around

DOI: 10.4324/9780429324901-2

half of the country's vote share between them, the Indian National Congress and the Bharatiya Janata Party, included several line items on digital technology in their manifestoes for the 2019 General Elections. This included the now common political promises like universal access to basic internet infrastructure and other niche subjects that intersect with e-governance, like data privacy and emerging technology (SFLC 2019).

At the face of it, India is successfully implementing its digital stack, if seen through a quantitative lens that is most easily applied to the infrastructural layer. For instance, the number of internet subscribers increased from a mere 148,000 in 1998 to 829.30 million as of December 31, 2021 (Telecommunication Regulatory Authority of India 2021). There are also over a billion mobile phone connections dotting the country's digital landscape (Telecommunication Regulatory Authority of India 2021). But, has the proliferation of digital technology in India improved the standards of living for its 1.3 billion strong population?

India ranked 129th out of 189 countries on the United Nations Development Programme's Human Development Index (HDI) in 2020 (Human Development Reports 2020).[1] Further, according to the Food and Agriculture Organisation, 14.5 percent of the country's population is undernourished.[2] And as per the latest round of the National Family Health Survey, more than half of the country's pregnant women are anaemic and child marriage of women remains unacceptably high (23.3 percent).[3] Several other developmental indicators point to a state of affairs that cannot possibly represent a swift march towards modernity or progress. We need not recount those glaring gaps in development outcomes here. It suffices to say that digital technology cannot drive a paradigm shift in development outcomes if it is used to stand in for the State rather than to make the State more effective. That is, digital is no substitute for a capacitated State. But why is e-governance, the long-hailed holy grail of administration, falling short of its transformational promise? Much like the green light calling to Mr. Gatsby in F. Scott Fitzgerald's seminal novel, Digital India continues to serve as a vision of "the orgastic future that year by year recedes before us" (Fitzgerald 2019). It may help to understand how digital technology has come to shoulder high expectations in the first place. A brief historical journey is in order to understand that the road to technological evangelism, which now manifests most prominently in a secular political affinity for digital, was paved with good intentions.

Nehru sets the stage for India's longstanding tryst with technology

India's interest in digitalisation and the services riding atop it stems primarily from a long prevalent political enthusiasm for technology. The country's first Prime Minister, Jawahar Lal Nehru, saw the development of technology as imperative for India's march towards modernity. According to Arnold (2013), in science and technology, Nehru found a banner under which he could reconcile ancient India's scientific achievements with the technological advancements garnered from interactions with Islamic and British civilisations (Arnold 2013). Science and technology offered Nehru a form of rationalist thinking that could prevent the violence threatened by the repetitive refrains of "religious bigotry and reactionary obscurantism" that pervaded post-colonial India (Arnold 2013).

Three notions bolstered Nehru's faith in science and technology in India (Arnold 2013). First, Nehru saw science and technology as the primary agents for socio-cultural transformation in the country (Arnold 2013). The proliferation of the "scientific temper" would root out an existing local mindset that viewed the country and its people through a hierarchical lens based on superstition and dogma (Arnold 2013). Science and the inventions afforded by scientific research would also expunge other prevalent social iniquities in India such as "hunger and poverty, insanity and illiteracy" (Arnold 2013).

Interestingly, after Nehru became more familiar with the intricacies of state planning and "rapid industrialisation", he began to talk more about the importance of technology rather than science (Sekhsaria 2018). The rhetorical shift is important, as it replaces the exploratory and broad scientific discipline with the most productive outcome of its course of discovery, namely technology. Evidentially, the Scientific Policy Resolution presented by Nehru to the Indian Parliament in 1958 provided that "the key to national prosperity" lay in the "effective combination of three factors, technology, raw materials and capital, of which the first is perhaps the most important, as the development of scientific techniques can make up for a deficiency in natural resources and reduce demands on capital" (Sekhsaria 2018). The shift was understandable to an extent. As a nation with widespread problems of a rather rudimentary nature, India could scarcely afford to direct resources towards scientific pursuits that may not fructify into something concrete. The higher purpose of science, and concomitantly that of technology, was to deliver the country from its most immediate woes.

Second, Nehru wanted the State to guide the direction of scientific and technological advances. At the time of independence, the British handed over several administrative matters to the many Indian states, including those pertaining to science and technology (Arnold 2013). However, science and technology became inextricably linked to independent India's notion of self-sufficiency, and thus, were deemed too important to be governed in a fragmented manner. Centralised control of scientific institutions and budgets also allowed Nehru to foster a "scientific establishment that was sympathetic to his views" (Arnold 2013). As a corollary, scientific research was to be carried out through institutional mandates. Nehru did not want scientific discovery to be the dominion of individual scientists carrying out experiments in ivory towers. Scientific research had to be geared towards broader national prosperity and modernisation, which in Nehru's mind could only be done if it was driven by the State (Arnold 2013).

Third, India's scientific and technological progress was directly proportional to the technical capability of its human resources (Arnold 2013). The 1958 Scientific Policy Resolution highlighted the importance of imparting proper scientific and technical training to as much of the population as possible (Sekhsaria 2018). Skilled human capital could then be leveraged to help the country forge its path forward. Specifically, quality technical human capital could be either used as a means of exchange for importing raw material or deployed towards helping create technologically advanced machinery that would help indigenous industries proliferate and assist the government with complex infrastructure projects (Arnold 2013). Thus, there was great emphasis on setting up institutes of scientific and technical training in India.

The first step towards creating a corps of proficient technicians for attaining this goal was to send Indian students abroad for technical training, for which the Government of India Fellowship Programme was established (Sharma 2009). Notably, one of the beneficiaries of this programme, Hargobind Khurana, won a Nobel Prize in Medicine in 1968 (Sharma 2009). The fellowship programme, however, was a stop-gap measure as it was not possible to scale capacity through this means alone. Plans were set in motion to create higher technical education institutes in the image of the Massachusetts Institute of Technology, one of the finest engineering colleges in the world (Sharma 2009). The first of these was set up in 1951 in Kharagpur, apparently because the town was home to India's "largest railway workshop" (Sharma 2009). It was originally named the "Eastern Higher Technical Institute" but later rechristened the Indian Institute of Technology (Sharma 2009). Over the course of the next few years,

IITs cropped up in all parts of the country such as Bombay, Madras, Kanpur and Delhi (Sharma 2009).

The Nehruvian vision for science and technology, particularly Nehru's belief that technology held the key to transforming India as a nation economically as well as socio-culturally, is critical to understand some of the factors underpinning India's current technological fervour in favour of digital. For one, successive governments similarly placed technology at the forefront of the national developmental imperative. Part of the reason may have been political or dynastic fealty, as many of the governments formed after Nehru's death, had his party, the Indian National Congress, at the helm.

Nehru's daughter, Indira Gandhi, who served as the country's prime minister from 1966 to 1977 and then again from 1980 to 1984, largely towed his line when it came to state-led technological development. During her first term, Indira emphasised self-reliance and the growth and development of domestic computer manufacturing capabilities, meaning strict restrictions on the import of technology (Sharma 2009). As a result, industry suffered, as did several prospective innovators wishing to try their hand at software development (Sharma 2009).

When Indira Gandhi came back to power in 1980, a change of heart manifested in some of her policies (Sharma 2009). A new electronics policy was announced which liberalised technology imports in a bid to help industry as well as those wishing to foray into software development (Sharma 2009). Indira's new liberal outlook towards technology was attributed to her son, Rajiv, who got increasingly involved in her political affairs right around the time she came back to power (Sharma 2009). Rajiv Gandhi was a technophile and surrounded himself with like-minded individuals whom he consulted frequently on matters of policy (Sharma 2009).

When India hosted the Asian Games in 1982, Rajiv worked personally with the National Informatics Centre to computerise operations surrounding the organisation of the Games, including the introduction of an electronic system that announced results automatically once an event ended (Sharma 2009). The Asian Games also revealed the first instance where technology was wielded to serve political interest under the garb of progressive policy. Prior to the games, India had taken an antagonistic approach to the personal audio-visual medium. Television ownership required a license and the import of televisions was highly restricted. This resulted in a deficit in television ownership around the time of the Games, which in turn, served as a damper in Rajiv Gandhi's plans to have the Games serve as the first colour telecast in the country (Sharma 2009). To see his vision achieve fruition, Rajiv pushed the government to liberalise television imports, despite opposition from the Department of Electronics (Sharma 2009).

Dynastic legacy notwithstanding, distinct political establishments such as the Bharatiya Janata Party, which subscribe to a political and cultural ideology that is completely divergent from that of the Nehru-Gandhi clan and the Indian National Congress, embrace technology with the same level of enthusiasm. Atal Bihari Vajpayee, for instance, a BJP stalwart who led the country as Prime Minister from 1999 to 2004, was instrumental in the induction of the National Telecom Policy, 1999 which placed great emphasis on access to telecommunications being necessary for the country to achieve its social and economic goals. Needless to say, telecommunications is the base for any digital communications.

Further on, Narendra Modi, the country's current Prime Minister who also hails from the BJP, is one of the most effusive champions of technology today. In a speech in 2017, PM Modi coined the equation, "IT + IT = IT", the expansion of which translates to Information Technology + Indian Talent = India Tomorrow (Singh 2017).

Why digital technology and politics make great bedfellows

There are several reasons why digital technology transcends political divides. First, is the institutionalisation of scientific and technological development in the country. Nehru had a direct hand in setting up and revamping most of India's main scientific research institutions (Arnold 2013). These entities worked closely with the government from their inception (Sharma 2009). For instance, the Council of Scientific and Industrial Research comes under the Ministry of Science and Technology, and its de jure head is the Prime Minister. The close nexus between the State and scientific research and education institutions fostered a culture of reliance between the two (Sharma 2009). Thus, scientists often had a say in the crafting of technology policy and the direction of technological development in the country (Sharma 2009). Now that a number of scientists are from a computer science background, digital technology policies are in sharp focus.

A second reason for technology's appeal to the country's political class is a strong track record of leading engineering institutions, which dates back to the early days of Independence. For instance, India's first analog computer was built at the Indian Statistical Institute (ISI) by Samarendra Kumar Mitra and Soumyendra Mohan Bose in 1953 (Sharma 2009). India also managed to develop an electronic switching

system for telecommunications soon after the first model was announced by Bell Laboratories in the US (Sharma 2009).[4] It was a remarkable achievement considering that only nine countries in the
world had this technology at the time. The success of the project
prompted the government to set work in motion for a 1000-line
switching exchange which was ready for commercial testing by 1980
(Sharma 2009).

The early Nehru dynasty's patronage of technological sectors coupled with the general success of attempts at innovation explains a third
reason why Indian politicians champion technology so emphatically
today, namely the development of a successful IT industry. The
overhaul of key technology sectors such as electronics under Indira
Gandhi, and telecommunications and computers under her son Rajiv,
served as a great boon (Sharma 2009). Rajiv Gandhi's 1984 computer
policy saw computer production in the country rise by 100 percent,
with a concomitant 50 percent drop in prices. India became a net
exporter of computer hardware (Sharma 2009). In a bid to bolster
software exports, a new policy was announced in 1986 that eased
foreign exchange rules for computer and software tool imports and
allowed foreign collaboration for software development (Sharma
2009). The telecom sector witnessed modernisation with the corpor-
atisation of telecom services (Sharma 2009).

Though common conception would have us believe that the push for
self-reliance under the Indira government stymied industry to a significant extent, it also precipitated the creation of a local electronics
ecosystem which also provided a bedrock for the growth of IT (Sharma
2009). Imports were strictly controlled and joint ventures with foreign
firms were generally forbidden (Sharma 2009). The latter was the fate of
Delhi Cloth Mills (DCM), a textile industry powerhouse, that wanted to
strike up a joint venture with Sony to manufacture calculators in India
(Sharma 2009). DCM wanted to make the move to electronics as the
commercial feasibility of textiles was wearing thin (Sharma 2009).
Despite the rejection of its application for a joint venture with Sony,
DCM went on to develop and launch the country's first domestically
manufactured desktop calculator in 1972 (Sharma 2009).

The local IT industry also benefitted from an early globalisation,
much before it was in vogue! The close involvement of the American
multinational – International Business Machines Corporation (IBM),
with the Indian government from the fifties to the seventies is hard to
overstate. IBM first came to India and set up shop with Nehru's assistance in 1951 (Sharma 2009). It bettered the fortunes of local computing
companies in two key ways. First, it nurtured a deep computer-centric

culture in the country (Sharma 2009). Over the course of 25 years, the company launched a concerted advocacy campaign, working closely with ministers, bureaucrats, other government officials – fostering a culture of governance centred around computers and high technology when there was little understanding about the utility of these machines in government (Sharma 2009). Additionally, IBM cultivated industrialists, and saw the introduction of computers in sectors such as "textiles, jute, and steel" (Sharma 2009). IBM also engendered a pool of highly trained computer and software engineers. It even went as far as training and educating academics and researchers and offered grants to educational institutions (Sharma 2009).

Industry also profited from the early investments in technical education that led to the emergence of a sizable contingent of highly trained personnel. The IITs incepted during Nehru were instrumental in the development of technical training in India (Sharma 2009). They were home to the first formal computer science education courses in the country (Sharma 2009). By the 1980s, all IITs were offering undergraduate, postgraduate and doctoral courses in computer science (Sharma 2009). Many of the individuals emerging from these courses went on to join India's burgeoning computer industry, enlisting with local firms such as DCM and Hindustan Computers Limited (HCL) (Sharma 2009).

Several other IIT graduates emigrated to America and went on to either found or lead American technology companies (Sharma 2009). Illustratively by 1998, Indian engineers were heading 775 tech companies in Silicon Valley (Sharma 2009). Notably, Junglee.com, founded by four IIT engineers, was purchased by Amazon for USD 160 million in 1998 – one of the largest such deals of the time (Sharma 2009). When India liberalised its telecom policy, several of the IIT engineers in leading positions in American companies convinced the latter to invest in India, engage in joint ventures with local firms, and set up subsidiaries in the country (Sharma 2009). One noteworthy joint venture was between HP computers and HCL, which saw the latter garner a global presence, something hitherto unheard of for Indian companies (Sharma 2009).

The combination of a favourable policy landscape and a wealth of well-trained and inexpensive manpower allowed Indian software exports to rise from INR 100 million in 1989 to INR 1 billion by 1991 (Sharma 2009). In July 1991, large chunks of the Indian economy were liberalised. A New Industrial Policy was introduced which eliminated licensing for all enterprises aside from those that directly affected public health and the environment (Sharma 2009). Public sector monopolies were done away with in most areas and foreign investment restrictions were removed. Liberalisation had an immediate impact on the domestic software industry; by 1992, there were 44 companies with sales of over INR 1 million, up from just 15 in 1991 (Sharma 2009).

The local IT services industry also saw great success during the Year 2000 or "Y2K" scare that cropped up during the mid-nineties (Sharma 2009). In 1959, Grace Murray Hopper and Robert Berner developed the Common Business Oriented Language (COBOL), a programming language that was ubiquitously used in the computing industry till the nineties (Sharma 2009). COBOL was devised to suit the needs of computers with limited memory capacities (Sharma 2009). To save memory, the format used to display dates on the COBOL computers used two digits each for the year, month, and day (Sharma 2009). So, 4th June 1991 would be displayed as 040691. The quandary here was that 91 could be 1991 or it could be 1691, and the inventors discovered the problem early on and inserted a modification that allowed the year to be written in four digits (Sharma 2009). However, when big computer companies such as IBM attuned COBOL to their own systems, they neglected to include the 4-digit year modification, continuing instead with the COBOL's original formulation (Sharma 2009). The computing industry persisted with the defect till the nineties when it dawned on users around the world that their systems may shut down on 1st January 2000, as their computers would display 00 for the year on that day (Sharma 2009).

Resultantly, several stakeholders began looking at ways to solve for the problem before it was too late (Sharma 2009). Rectifying the COBOL issue would be costly and manpower-intensive, as each faulty programme had to be manually identified and rewritten (Sharma 2009). As the US and Europe did not have the technical labour-force to handle this workload, they looked to other geographies for resolution (Sharma 2009). India was a prime choice – with its vast contingent of inexpensive engineers that spoke English reasonably well (Sharma 2009).

The Indian government moved to enable its private sector to capitalise on the Y2K opportunity. The Department of Electronics published a report on the Y2K problem and pushed government entities to proffer "courses on COBOL and the Y2K bug" (Sharma 2009). A mass marketing effort was launched to signal to firms in the US, Japan, and Europe that India was the place for Y2K solutions (Sharma 2009). Special incentives were designed to stimulate interest amidst local industry in India (Sharma 2009). Most large to medium-sized firms partook and reaped benefits for doing so. Between 1996 and 1999, Indian software companies earned USD 2.3 billion from

Y2K-related work carried out for American and European companies (Sharma 2009).

As the prowess of the IT industry grew, India placed greater faith in the ability of digital technology to turn the tides of its socio-economic fate. The Tenth Five-Year Plan, which charted out India's economic roadmap from 2002 to 2007, emphasised the potential of IT to bridge gaps between regions and classes of people, improve the state of governance, and help small businesses, in addition to looking at ways to consolidate the country's success in IT exports further (Planning Commission 2002). Importantly, IT features in nearly all departmental outlays given in the Tenth Five-Year Plan, either in the form of a scheme to be floated by a ministry, or for "modernisation" through computerisation and other process upgradation through technology (Planning Commission 2002). The Eleventh Five-Year Plan (2007–2012) doubled down further on the IT-centric planning of the Tenth Plan, with the Department of Information Technology, the nodal government agency for all matters pertaining to information technology in India, seeing a 498 percent increase in its budgetary outlay over the previous plan period (Planning Commission 2007). The Twelfth Five-Year Plan, 2012–2017, proffered more of the same, although there was a decreased emphasis on the IT industry (possibly as the latter took a significant blow after the 2008 financial crisis). Developing domestic human capacities through technology and encouraging technological innovation, however, were top on the agenda.

India's development policy planning from the early 2000s closely mirrors narratives floated by Bretton Woods Institutions like the World Bank that tout digitalisation as a catalyst for rapid socio-economic advancement in developing nations. They would have us believe that digitalisation can usher in socio-economic paradigm shifts by overcoming physical limitations, such as the absence of quality roads, through a global communications superhighway that grants access to information. Illustratively, a United Nations Development Program's 1999 Human Development Report cites how digitalisation could benefit the poor (Brown 1999). It suggests that digitalisation may bridge critical knowledge gaps in "information poor hospitals and schools" and empower marginalised minorities by allowing them to organise online (Brown 1999). The Report also notes how digitalisation would also potentially change the fortunes of small businesses by enabling them to optimise existing resources and connecting them to distant, more lucrative markets (Brown 1999). Finally, it discusses the role of digitalisation could play in transforming governance by allowing for more data-centric policy design (Brown 1999). While some of these contentions may hold, in part, they require qualification.

When the Modi government came to power in 2014, the Planning Commission, the apex body for all economic planning in the country, was disbanded and replaced by a new body, the National Institution for Transforming India (NITI) Aayog. The acronym NITI is the Hindi term for "policy". This was a signal that India wanted to evolve from a centrally planned economy to one where private industry had greater reckoning. The NITI Aayog was fashioned more in the form of a think tank that would aid both the Centre and states in matters of policy.

Since its inception, the NITI Aayog has come out with several strategy papers that centre around the transformative potential of new technologies. Most recently, there was a paper on mobility that discussed the benefits of ridesharing applications and an Artificial Intelligence (AI) Strategy Paper that elaborates upon the hefty economic promise of the large-scale deployment of AI in India (Niti Aayog 2018; 2021). NITI's most comprehensive policy document, the Strategy for New India at 75, discusses the key areas the government must focus on to chart the course for a new, transformed India by 2022 (the country's 75th year of independence) (Niti Aayog 2018). Here again, there is quite a bit of emphasis on technology as a key driver for economic growth and a solution for multiple societal woes.

India moved from strength to strength on the back of its technological achievements, and it makes sense to put political weight behind technology. It managed to reengineer, often in the face of great adversity, complicated high technology. When the West, typically considered far ahead of the South Asian subcontinent in technical matters, stood stymied by technological problems such as in the Y2K episode, India stepped in and not only sorted the problem out, but built one of the world's most successful information technology export industries on the back of it. India currently boasts of the second largest internet userbase in the world and is ostensibly primed to reap digital dividends, or so it is suggested by several reports produced by both development agencies and government.

A politician that champions technology is therefore seen as an individual concerned with "development and modernisation" (Sekhsaria 2018). Talking up the importance of technology could endear such a politician to people from across social and economic strata. The current Prime Minister, Narendra Modi, is seen as a reformer by many because of his emphasis on e-governance and digitalisation – which is most visible in his government's Digital India scheme and vision. Digitalisation presents itself as that missing link whose arrival portends benefits for small businesses, marginalised communities, and education and health centres in underdeveloped areas. The pliability of technology, in turn, consolidates its position as a socio-culturally and economically neutral phenomenon. Technology can and will work for anyone. It can be applied across different socio-economic contexts.

Emerging technology is also generally esoteric, an enigma whose mystery lends itself well to its promises of prosperity. The unknown provides a blank canvas that may be coloured with hope, especially by the desperate and the destitute. The esoterisms of technology make it an effective political platform, ripe for marketing. The combination of these factors is instrumental in aiding the spread of digital technology in India, allowing politicians to take the credit for incremental adoption, with varying degrees of impact on socio-economic progress.

Technology as the only means to navigate structural deficits

The Indian State is short-handed (a dearth of capital as well as qualified, capable, and willing individuals) and therefore ill-equipped to deal with the severe developmental problems plaguing the country. Technology, then, serves as a sort of bionic prosthetic, one that not only replaces missing elements in government and governance but also enhances national capability to deal with such challenges. Cars and airplanes help us reach further than our legs will take us in a given period. Telephones carry our voices to those well beyond shouting distance. For the poor farmer, engineered seeds and chemical fertilisers ensure them a larger yield, pestilence proof and hassle-free. For government, digital technology serves as a means to reach those remote areas that remain beyond the reach of the state. Rather than roads, the State reaches its citizens through digital highways.

Modernity is defined by a kind of sterility that has worked to slowly but surely move organic life out of the realm of human existence and each technological innovation driving it amplifies this decontamination. If we understand modernity to be an evolution of humanity from a primitive, nature-bound species to an urbanised sophisticate, we can look to technology serving as the primary means of such transformation. And with each technological wave, the messiness of humanity, of being, of life finds itself obsolete or hindering the march of progress. A living being is unpredictable, messy and inadequate in the face of the machine. Machines march on tirelessly, and are, therefore, presented as the best versions of organic beings. Thus, human-driven cars replaced human-guided horse carriages, and if all goes well the human will be booted out of the driving equation completely with autonomous vehicles.

Each iteration of technology presents an improved version of its prior avatar. And each wave of technological evolution brings with it an exponential increase in its abilities to solve our problems autonomously, to enhance our abilities to the point that humans are no longer part of the equation. Digital technology is the latest innovation on which we have pegged such hopes. It is purported to solve a range of India's developmental woes. However, the utility of this technology is directly proportional to the way or the context in which it is deployed.

At this stage, it's also worth noting that contradictions surrounding digital technology are not novel. Indeed, paradoxes have cropped up through the last century with a sundry of different technological advances. Lukasiewicz (1994), for instanced pointed out that the technological advancements that spurred industrialisation also precipitated a transformation of our physical surroundings into more urbanised settings. The structure of society evolved concomitantly, shifting from "a relatively static form of social organization" to an increasingly complex and dynamic paradigm (Lukasiewicz 1994). These changes were characterised as a march towards civilisation and progress. In this predominantly occidental view, progress meant increases in material wealth, the spread of "social services, and an emphasis on the societal procreation of knowledge and information to ensure a continuous churn of technological advancement across the fields of human industry" (Lukasiewicz 1994). Lukasiewicz (1994), however, noted that such a notion of progress was rather limited in scope, as it failed to account for the contradictions that arose from humankind's pursuit of it. Specifically, the intellectual capabilities and resultant technological innovation that afforded humankind the ability to transform and exploit its physical environment failed to equip us with the ability to adequately manage the latter's newfound artificial edifice. Lukasiewicz (1994) cites the inability of humankind to deal with climate change and control the adverse effects of pollution as evidence of this incongruity.

Astrophysicist Carl Sagan was probably one of the most outspoken commentators on the inconsistencies between our expectations from technology and the manifest outcomes of its spread. In Pale Blue Dot, Sagan (1997) points out how humanity's attempts at averting crises through technological intervention, often lead us to "usher in" graver ones. One example he cites is that of ChloroFluorocarbons (CFCs), compounds that were expressly developed as a safe alternative to existing refrigeration chemicals such as ammonia that tended to leak and were potentially life-threatening if ingested (Sagan 1997). CFCs were inert gases that were odourless, tasteless and colourless (Sagan 1997). The inertness of CFCs ostensibly rendered them innocuous, and thus found them at the centre of many practical applications beyond re-frigeration, namely as propellant for aerosol cans and packing material (Sagan 1997). However, Sagan points out that the scientists that developed them failed to identify a glaring caveat to the so-called in-vulnerability of CFCs (Sagan 1997). When these molecules reached the Earth's stratosphere, they would be broken down by the Sun's ultra-violet light, releasing chlorine atoms that would bind with the ozone layer, thereby eroding it (Sagan 1997). CFCs, then, went from a technological development targeted at safeguarding human beings to become key contributors to one of the most severe existential threats facing our species – Global Warming (Sagan 1997).

India is no stranger to unintended consequences of the use of technology out of context, as a substitute for the State. For instance, the push for population control through medical advances in the 60s and 70s, ostensibly to bolster economic growth by ensuring that the finite resources in the country are not overwhelmed by population growth, led to a skewed sex ratio (Hvistendahl 2011). The country's birthrate in the mid-sixties was 5.7 children per woman. A group of medical scientists, sent to India at the behest of the US government and a handful of aid agencies, discovered that Indian families are disproportionately large because parents kept conceiving until they had a son. For a long time in India, the property was only inheritable by male offspring. Even now, certain religious rites can only be carried out by men.

On discovering that the desire for male progeny lay at the heart of India's booming population, American doctors stationed at the All India Institute of Medical Sciences (AIIMS) began introducing sex determination techniques such as amniocentesis. AIIMS was a seedbed for leading physicians and public health officials in India; positioning a contingent of advisers there gave Western aid agencies "access to both doctors as well as policymakers" (Hvistendahl 2011).

AIIMS began its amniocentesis trials in 1975 with the official line being that it was to detect fetal abnormalities (Hvistendahl 2011). Shortly thereafter, however, a number of AIIMS doctors collectively published a paper that candidly explained what the tests were actually being used for (Hvistendahl 2011). They noted that socio-economic dynamics prompted Indian couples to prefer having a male child and this desire generally compelled them to reproduce until a son was born. The doctors went on to say that sex determination did away with such "unnecessary fecundity" (Hvistendahl 2011).

Other public hospitals began offering amniocentesis tests soon after AIIMS began its trials (Hvistendahl 2011). Although the government banned sex determination in public hospitals shortly thereafter, the seeds for sex-selective abortion were already sown (Hvistendahl 2011). The practice quickly migrated to private clinics. An official inquiry on the amniocentesis procedure revealed that between 1978 and 1983, approximately 78,000 female fetuses were aborted in India (Hvistendahl 2011). In 1981, the ratio of women to men in India, the gender ratio, was 934 to 1000. By 1991, it had dropped to 927 females per 1000 males (Hvistendahl 2011). Sex determination was banned in 1994 but it had little effect on improving the sorry state of affairs in the country (Hvistendahl 2011). In 2001, India's sex ratio was still only 933 females per 1000 males, with a greater skew in urbanised states like Delhi (Hvistendahl 2011). Although efforts have been made to stem the tide of sex-selective abortions and the accompanying practice of female infanticide, reports indicate that both remain rampant.

Amniocentesis was developed to detect fetal abnormalities but was used in India to aid technology-enabled genocide of female babies. This unintended consequence stemmed from a lack of state capacity to educate citizens and to hold private clinics to account. The State abdicated its responsibility to deliver public goods like family planning education, whereas the introduction of a new technology, even though completely out of context, was welcomed by an uninformed citizenry.

We would like to say at this point that we are not techno-antagonists. We readily acknowledge digital technology proffers many benefits to those who are able to use it. Its utility as a tool for spreading awareness is exemplified by India's experience with the COVID pandemic. Uohara et al. (2020) note that technology-enabled remote patient care facilitated contact tracing, and created awareness about safety protocols such as wearing masks and social distancing. Technology enabled children to continue with schooling and businesses to function remotely. Our own experience with the pandemic saw how citizens came together to help each other find resources such as oxygen and hospital beds in the wake of the second wave of the pandemic. Without technology, that vast effort of citizen initiative would not have been possible. However, technology was not the ultimate solution but only part of it. The effort involved individuals verifying the availability of resources and updating databases of useful information. In the context of state interventions that utilise technology, however, its utility is also exaggerated to the point of caricature.

Digital technology is often looked upon as *the* solution which allows politicians to avoid dealing with the intricacies or messiness of governance. That is, digital technology is used as a substitute rather than a complement. This paradigm is typified by the myriad digital governance initiatives aimed at empowering communities most dependent on state support. We detail this assertion with examples in the subsequent chapters.

The goal of this book is to prompt Indians to rethink the solutionist mindset that is so pervasive in political discourse on technology today. Development is seldom easy to reverse engineer through the use of technology. True progress involves wrestling with the tricky problems that pervade markets and society. Smart governance demands melding the physical State and digital technology's promise together to unlock developmental benefits. In short, technology stacks are not solutions in and of themselves. They are not meant to stand in for the State. They are means for governments to respond to developmental challenges, when employed in context-specific ways, as we further discover in subsequent chapters.

Notes

1 The HDI measures development progress on several parameters that broadly evaluate the average state of health (life expectancy), knowledge (average years spent schooling), and standard of living (Gross National Income) in a country.
2 FAO, IFAD, UNICEF, WFP and WHO. 2019. The State of Food Security and Nutrition in the World 2019. Safeguarding against economic slowdowns and downturns. Rome, FAO. http://www.fao.org/3/ca5162en/ca5162en.pdf.
3 National Family Health Survey, 2019–21.
4 Fundamentally, a switching system allows a wireline telecommunications network to optimise its transmission lines so that a separate line does not have to be created for every individual with a telephone (Britannica 2017).

Bibliography

"A Look at Party Manifestos for the 17th Lok Sabha Elections." Software Freedom Law Centre, 2019. https://sflc.in/look-party-manifestos-17th-lok-sabha-elections-will-political-parties-defend-our-digital-freedom.
Adams, Douglas. *The Hitchhiker's Guide to the Galaxy*. Random House Publishing Group, 1989.
Arnold, David. "Nehruvian Science and Postcolonial India." *Isis* 104, no. 2 (June 2013): 360–370, https://www.journals.uchicago.edu/doi/pdfplus/10.1086/670954.

Bhandari, Apurva, et al. "Moving Forward Together: Enabling Shared Mobility in India." New Delhi: Niti Aayog, 2018. https://www.niti.gov.in/writereaddata/files/document_publication/Shared-mobility.pdf.

Britannica, T. Editors of Encyclopaedia. "Switching." Encyclopedia Britannica, August 3, 2017. https://www.britannica.com/technology/switching.

Brown, Mark. "Human Development Report 1999." New York: United Nations Development Programme, 1999. http://hdr.undp.org/sites/default/files/reports/260/hdr_1999_en_nostats.pdf.

Chatterjee, Shoumitro, and Mekhala Krishnamurthy. "Understanding and Misunderstanding E-NAM." *Seminar*, January 2019. https://www.india-seminar.com/2020/725/725_shoumitro_and_mekhala.htm.

Commission, Planning. "Tenth Five Year Plan 2002 – 2007: Volume II Sectoral Policies and Programmes." Government of India, 2002. https://niti.gov.in/planningcommission.gov.in/docs/plans/planrel/fiveyr/10th/10defaultchap.htm.

Commission, Planning. "Eleventh Five Year Plan 2007 – 2012." Government of India, 2007. https://niti.gov.in/planningcommission.gov.in/docs/plans/planrel/fiveyr/11th/11_v1/11th_vol1.pdf.

Haq, Zia. "Jal Jeevan: 51 Million Rural Households Now Have Tap Water, According to Govt Data." *Hindustan Times*. August 19, 2020. https://www.hindustantimes.com/india-news/jal-jeevan-51-million-rural-households-now-have-tap-water-according-to-govt-data/story-JepJ70SGthLh66bOkuO3bJ.html.

Fitzgerald, F. Scott (2019). The Great Gatsby. Wordsworth Collectoras Editions.

"Human Development Reports." United Nations Development Programme, 2020. http://hdr.undp.org/en/countries/profiles/IND#.

Hvistendahl, Mara. *Unnatural Selection: Choosing Boys Over Girls, and the Consequences of a World Full of Men.* New York: Public Affairs, 2011.

Lukasiewicz, Julius. *The Ignorance Explosion: Understanding Industrial Civilization.* Ottawa, Canada: Carleton University Press, 1994.

"National Strategy On Artificial Intelligence." Niti Aayog, 2018. http://niti.gov.in/national-strategy-artificial-intelligence.

Sagan, Carl. *Pale Blue Dot: A Vision of the Human Future in Space.* Penguin Randomhouse, 1997.

Satish, Rohit, and Tanay Mahindra. "Responsible AI #AI For All: Approach Document for India Part I - Principles for Responsible AI." New Delhi: Niti Aayog, February 2021. https://www.niti.gov.in/sites/default/files/2021-02/Responsible-AI-22022021.pdf.

Sekhsaria, Pankaj. *Instrumental Lives: An Intimate Biography of an Indian Laboratory.* Routledge Focus on Modern Subjects. Routledge, 2018.

Sharma, Dinesh. *The Long Revolution: The Birth and Growth of India's IT Industry.* India: Harper Collins Publishers, 2009.

Sharma, Ram Sewak. "Digital India: A Programme to Transform India into a Digitally Empowered Society and Knowledge Economy." Presented at the

Digital India: Transforming India into a Digitally Empowered Smart Nation, Washington D.C., October 15, 2015. https://www.meity.gov.in/sites/upload_files/dit/files/Digital%20India.pdf.

Singh, Soibam Rocky. "PM Modi's New Formula for India's Future: IT+IT= IT." *Hindustan Times*. May 10, 2017. https://www.hindustantimes.com/india-news/sc-to-go-paperless-pm-launches-integrated-case-management-system/story-d5RviRBMUZ5J5w4KDkkn7I.html.

Sridhar, Vardharajan. *The Telecom Revolution in India: Technology, Regulation, and Policy*. New Delhi: Oxford University Press, 2012.

"The Indian Telecom Services Performance Indicators April – June, 2021." New Delhi: Telecom Regulatory Authority of India, October 21, 2021. https://www.trai.gov.in/sites/default/files/PIR_21102021_0.pdf.

Uohara, Michael, James Weinstein, and David Rhew. "The Essential Role of Technology in the Public Health Battle Against COVID-19." *Population Health Management* 23, no. 5 (October 5, 2020). 10.1089/pop.2020.0187.

Yadunath, Darshan. "Transforming India through Digital Innovation." *Digital Development - World Bank Blogs* (blog), March 28, 2016. https://blogs.worldbank.org/digital-development/transforming-india-through-digital-innovation.

3 Defaulting to e-Governance

The idea that digital technology enhances governance capabilities is not new. Scholars have argued that it enhances efficiency and transparency (Finger and Pecoud 2003; Heeks 2001), extends the hours of service availability (Saxena 2005), increases accountability, enhances development potential and can be used to empower citizens (Heeks 2001). However, critiques of such stances are not new either. Heeks (2001) has noted that most digital governance schemes end in failure and so-called "success stories" tend to be the exception rather than the rule. The stories of failure ring particularly true for developing nations in these accounts (Heeks 2001).

Yet in India, digital technology has never featured more prominently in the governance apparatus. The country exemplifies digitally mediated governance. We find the digital percolating into every ambit of the country's development schemes. The telecom revolution has become a digital one. The State is continually looking for ways to leverage India's billion mobile phone connections and hundreds of million internet users for its schemes.

This chapter explores a central puzzle of e-governance interventions, namely the State's unwavering commitment to them despite a limited track record of success. To illustrate this puzzle, we explore digital interventions targeted at solving for two long-standing development concerns – agricultural extension and land records. Both are complex issues that have found no ready resolution over the years (Chatterjee and Krishnamurthy 2019; Wahi 2016).

Several agricultural extension initiatives in the country have ended in failure, either as a consequence of inadequate calibration of a scheme to local context or insufficient financial support. Schemes that have enjoyed success were tailored to farmers' needs and closely monitored by research institutions (Satyanarayana et al. 2006; Pandian et al. 2014). However, in a bid to scale agricultural extension

DOI: 10.4324/9780429324901-3

efforts, without commensurate increase in resource outlays, the State began introducing digital extension programmes. We use the example of two such programmes, one targeting the pre-production phase in agriculture known as "Mkisan", and the other geared towards improving the state of agricultural marketing, "E-Nam".

Similarly, land records[1] in India are in a state of disarray. Land registries are riddled with problems of non-standardised and outdated information, backlogs and corruption (Bal 2017). The uncertainty around land titles has led to property being the most litigious issue in the country. According to one survey, 67 percent of cases in the country relate to property (Narasappa et al. 2016). The National Land Record Modernization Programme was introduced in 2008 to resolve some of these issues (Deshpande 2007). However, the programme continues to be a non-starter in many ways (Shah et al. 2017). Despite the failure of the Land Record Modernization Programme, however, we find that the State seeks to extend the scope of its operations, without addressing the flaws in its design.

In the first part, we unbundle the puzzles surrounding technologically mediated governance and development schemes. Though digital proffers interactivity, we find that it often works to enhance the centralisation of governance schemes. Schemes that are not adequately localised fail because they do not address problems on the ground or do not secure buy-ins from stakeholders at this level. Next, we find that while digital technology is often looked at as a means to enhance state capacity, its inclusion in governance schemes works to erode it.

Subsequently, we assess the State's attempts to wield blockchain technology to add yet another technological layer to resolve the problem with land records. Like all digital technology, blockchain and decentralised ledgers are an emerging technological construct that have the potential to vastly improve governance. The utility of blockchains lies in the creation of immutable records through a combination of cryptographic techniques. As such, their role in solving various problems in governance is currently under consideration. Foremost amongst such prospects is the use of blockchain to solve the problems of land records. We highlight why this may make heighten the complexity of existing challenges, in the unique Indian context.

Digital makes governance harder

There are several puzzles that emanate from the rollout of technical solutions for governance in India. First, digitally mediated governance programs tend to exacerbate a top-down governance approach, despite their ability to facilitate effective decentralisation through feedback mechanisms.

The inability to adequately localise solutions was a key driver of the failure of past agricultural extension programmes in India, as exemplified by the Train and Visit programme. T&V involved training programmes run by subject matter specialists, analogous to a "train the trainers" mode of capacity building (Sajesh and Suresh 2016). Although T&V was initially successful and scaled up over the 20 years after it was incepted, it had several shortcomings. The primary drawback of T&V was that it was a supply driven programme, where information trickled down from research institutions to farmers with little incorporation of farmer feedback (Sajesh and Suresh 2016). Thus, it was often the case that the information shared with farmers was of little relevance to them in the long run. Consequently, the programme ended at the beginning of the nineties.

The problem of over-centralisation due to top-down design even percolated into extension programmes that were otherwise localised to work with recipients, as in the case of the Agricultural Technology Management Agency (ATMA). The ATMA was designed to align public extension programmes with market realities. It aimed to facilitate greater decentralisation of extension activities, allowing for better linkage between farmers and research institutions (Tamil Nadu Agricultural University 2015). Each district had an ATMA registered as a society (Tamil Nadu Agricultural University 2015). Membership comprised of all key stakeholders of the agricultural ecosystem, increasing accountability to the farming community (Tamil Nadu Agricultural University 2015). Sulaiman (2012) notes that while the ATMA pilot was successful in seven states, namely Himachal Pradesh, Punjab, Bihar, Jharkhand, Orissa, Maharashtra, and Andhra Pradesh (Singh et al. 2013), the broader nationwide rollout was not. Aside from problems of capacity, Sulaiman and Hall (2008) note that as ATMA was a centrally promoted scheme, there was a "lack of local ownership". State extension bodies saw ATMA as "just another central scheme they have to implement".

Digitisation engenders the potential for an interactive extension service, something remote physical extension programmes are unable to offer. As Sulaiman and Hall (2008) note, the information needs of farmers are wide-ranging and constantly changing. Farmers needed information on "not just technology" but also "on prices, consumer preferences, markets, and trade regulation" (Sulaiman and Hall 2008). Such information needs required "rapid responses and solutions"

(Sulaiman and Hall 2008). The dynamism, speed, and interactivity afforded by digitalisation seemingly meet these requirements.

However, digitisation carried forward and, indeed, exacerbated the design flaws of past extension programs, as illustrated by two schemes, one related to production stage of agriculture called "MKisan", and the other related to the post-production stage called Electronic National Agricultural Market or "E-Nam". Like the T&V, both Mkisan and E-nam follow a top-down topology. Despite their digital backbone, they operated as one-way digital portals and have limited scope for recipient feedback. On Mkisan, for instance, when text messages are sent out, there is no option for a farmer to respond to them (Afroz et al. 2018). According to available research, foremost amongst farmer grievances with Mkisan is the difficulty in resolving uncertainties about the information shared (Afroz et al. 2018).

Similarly, E-Nam merely broadcasts prices of different commodities rather than offers a practical means of direct online trade for farmers. The portal does not offer farmers an end-to-end digital commerce option. They cannot sell their produce to traders directly through it. They must instead visit the APMC, where commission agents mediate their interaction with the portal and assist them with participation in an auction. Commission agents are powerful intermediaries that form an important part of the farmer agency problem. In states with APMCs, these intermediaries enjoy a great deal of power because they have a monopsony on agricultural purchases. Although there is now an E-Nam mobile application that grants the farmer greater autonomy to manage a trade, there is no data on how many farmers downloaded the application and used it for transactions. Even with the mobile application, however, the farmer must still come to the mandi premises to engage in trade, which defeats the convenience element of online transactions.

For offline interventions, the barrier against successful localisation is largely a function of state capacity. The capacity deficits in digital interventions, however, have more complex dimensions, which brings us to the second puzzle surrounding their introduction. On the one hand, digital schemes are viewed by the State as a means of enhancing capacity. However, we observe that more often than not, digital technology creates its own set of capacity problems. From the vantage point of the stakeholders on the ground such as the farmer or the village-level land administrator, capacity relates to their ability to harness technology to their benefit. Although digital is often presented as an inclusive medium, it is by its very nature exclusive. Using digital technology requires a modicum of literacy on the part of the user.

The more technical the technology, the greater the demand on the intellectual capability of the user. Consequently, deficiencies in cognitive capacity or technical understanding may operate as a barrier towards technology adoption.

Cognitive capacity served as a barrier for the schemes discussed in the present chapter. For instance, many village-level land administrators (the officers responsible for entering details into the land registries) were unfamiliar with the software used for the Digital India Land Records Modernization Programme (Shah et al. 2017). Similarly, a survey of literacy amongst the farming community in India (Table 3.2), indicates that both Mkisan and E-nam would have limited utility for a majority of them.

From the vantage point of governance, the clearest manifestation of state capacity deficits is poor design of a digital scheme. Good design involves understanding the breadth of a governance problem, evaluating how technology can be helpful in resolving it, appreciating the limitations of such an intervention, and inducting supplements to account for such limitations. Such assessments are generally absent when a digital scheme is introduced, evinced by the fact that these programmes are usually introduced to solve hard problems that do not have straightforward solutions. The more unwieldy a governance problem, the greater the temptation on the part of the State to streamline it with technology. Such interventions

Table 3.1 Education Levels Among Agricultural Workers (%)

Education level	Share of Workers in Agriculture (in the survey sample)	Share of Workers in Agriculture (in the estimated number of workers employed in the agricultural sector in India)
Not Literate	**25.01**	**31.12**
Literate and up to Primary	17.33	18.74
Middle	17.81	16.44
Secondary	10.13	8.27
Higher Secondary	6.06	5.17
Diploma/ Certificate Course	0.36	0.33
Graduate	2.99	2.64
Post Graduate and Above	0.49	0.43
Secondary and Above	19.83	16.85

Source: Annual Report of the Periodic Labour Force Survey 2018–19.
Notes: Calculated using Table 21: Percentage distribution of persons by usual status (ps+ss) for each general educational level. (rural + urban) (persons).

hint at a presupposition that the introduction of technology in an informal and chaotic ecosystem riddled with market failures will somehow smoothen these irregularities. The consequence of such an outlook towards digital interventions is that once the State introduces them, it typically does not work to overhaul them, even in the event of failure. Rather, it seeks to extend its scope or add another layer of technology to rectify the problem. There lies the third puzzle of the digital. It is generally introduced to reduce complexity in a system but often serves to exacerbate it.

In the case of land records, computerisation has been presented as a solution since the time of the Seventh Five-Year Plan (1985-90) (Deshpande 2007). Under this Plan, the Centre launched a scheme to computerise land records in several districts across nine states in the country (Deshpande 2007). Its objectives were to computerise ownership details so that updated copies could be shared with landholders and create a cost-effective and reliable land records storage system that facilitates easy retrievability of information. Under the Eighth Plan, the Computerisation of Land Records (CLR) scheme was to be extended to 323 districts and 177 more in the subsequent planning period (Deshpande 2007).

However, the CLR scheme made little headway (Deshpande 2007). Problems included delays in fund disbursal, software development, and technical capacity limitations among government staff (Deshpande 2007). Moreover, computerisation did not involve rectifying incorrect records (which would be done through a survey) (Deshpande 2007).

Despite these issues, the Government embarked on a land record modernisation programme in 2009. This integrated the CLR with the Ministry of Rural Development's Strengthening of Revenue Administration (SRA) and Updating of Land Records (ULR) schemes, which had been launched by the Ministry of Rural Affairs and Development prior to the advent of the CLR (Nayak 2013). The new programme was known as the National Land Record Modernization Programme (NLRMP). Broadly, the objectives of the NLRMP were to modernise the system of land records, reduce property-related litigation, facilitate transparency in the upkeep of land records. The idea was to eventually move towards an immutable land titling program which unified the institutional interface for land-record management and ensured fidelity between property ledgers and the ground realities surrounding ownership (Nayak 2013). Unlike the CLR which focussed largely on technical elements of record keeping the NLRMP also included amendments to various legislations and capacity-building of staff as part of its scope (Nayak 2013).

The NLRMP set an ambitious target of computerising all land records in the country by 2017, through a public-private partnership model (Bal 2017). It did not meet its objective, largely due to the prohibitive costs of the project. There were reports of outdated records and corruption. The NLRMP was promptly re-christened in 2016 as the Digital India Land Records Modernization Programme (DILRMP) (Bal 2017). The objective and scope of the DILRMP was largely the same as the NLRMP except that it also incorporated the use of additional technologies such as a Geospatial Information System (Bal 2017).

A review of the implementation of the DILRMP in Rajasthan, however, found several issues in its operations. By 2017, Rajasthan had successfully digitised the Record of Rights or the RoR for 46,000 out of 47,918 villages (Shah et al. 2017). However, data in the RoR was non-standardised and the encumbrances on the property (i.e. if it is disputed or mortgaged) was not recorded (Shah et al. 2017). Moreover, legacy records were not available online, which meant that any due diligence on a property transaction would require an inspection of physical records (Shah et al. 2017). Further, the absence of legacy records meant that officers do not have any convenient verification mechanism, should the need arise (Shah et al. 2017). Though maps had not been updated through "modern survey methods", existing cadastral maps were scanned and available online and the digitisation of the process of registering property transaction documents had progressed satisfactorily (Shah et al. 2017). But, the primary database for cadastral maps could only be updated every 20 years as per an existing legal provision (Shah et al. 2017). During the interim period, these maps were being updated by hand and then scanned for the digitisation process (Shah et al. 2017). Not only did digitisation consolidate the errors in the handmade maps but it was also redundant as the land was being re-surveyed (Shah et al. 2017). Lastly, the registration process for land transaction documents did not involve any verification of title (Shah et al. 2017). Thus, even illegal property transfers were being stored in a digital ledger.

It is likely that these problems found in the implementation of DILRMP manifest across the country. Yet, the Government is now looking to extend the scope of the programme further, by introducing the Unique Land Parcel Identification Number (ULPIN), a 14-digit alpha-numeric code that will be assigned to every land parcel in the country. ULPIN will be launched across ten states in the country, and its rollout is expected to be completed by 2022 (Jebraj 2021). The government also announced that Aadhaar cards will eventually be linked to ULPINs to mitigate fraud (Jebraj 2021).

Aside from importing the myriad problems with Aadhaar-based authentication into land record management, which are highlighted in Chapter 5, scholars contend that it is likely that ULPIN will compound existing problems with land records if legacy issues are not addressed (NDTV Gadgets 360). The Government is carrying out large-scale surveys to ensure property records and maps are up-to-date. However, the absence of wholescale legal and administrative reforms in land record management indicates that even intensive survey endeavours will not rectify problems with land records on the ground.

The stratification of broken digital programmes into multiple technological layers brings us to our fourth puzzle, namely the use of digital technology to scale government service delivery. Digital technology is seen as a means to scale up development programmes, to democratise their benefits. However, the downside of such wide-ranging democratisation is lowest common denominator service provision with minimal potential for impact. To recap, Mkisan, provides high-level advisories and encompasses no mechanism for recipient feedback or clarification (Figure 3.1).

Likewise, information on E-Nam is often outdated or incomplete (Chatterjee and Krishnamurthy 2019; Dalwai 2017). Price data posted on the E-Nam portal is often sourced from offline auctions or government procurement (Chatterjee and Krishnamurthy 2019). Mandis avoid putting "high-volume" commodities through the system because the auction process takes time (Chatterjee and Krishnamurthy 2019; Dalwai 2017).

Mr. DR. SUKANTA BISWAS	Senior scientist and Head	26 Apr 2018		If the spray of urea at 10 percent of the jute land, the jute tree will grow rapidly and the weed reduction will be less.
DR. D SEKHAR	Technical Officer	26 May 2020		Farmers are advised to go for soil sampling for soil test, that helps in crop management as per soil fertility status in kharif season.

Figure 3.1 Example of Advisories Sent on Mkisan.

The other incentive for the State to look at digital interventions as a means to scale is the prospect of cost saving. The notion that digital interventions are cost-effective perhaps emanates from the belief that they involve less physical heavy lifting, especially if the backbone connectivity infrastructure is in place. Indeed, over the past few years, India has seen a rapid growth in domestic mobile and internet services provision. While there are still gaps in access to information infrastructure, the country is much further along on its journey than others that are similarly placed socio-economically, at least in terms of the ubiquity of mobile internet services (Table 3.2). The concomitant idea of digital abundance, in turn, prompts belief in the State that the digital realm is a ready resource that may be tapped for cost-effective development interventions.

The "ready resource" contention linked to connectivity infrastructure is, however, a dangerous misconception as it overlooks the myriad confounding factors that impede development interventions. These include the cost of encouraging uptake of technology amongst new users, both on the supply and demand sides, including the cost of subsidy. Even if governments do not charge for public services, the telecom and internet carriers most certainly do. For instance, local rates apply to many of the text-based services on Mkisan. And existing research suggests that there is a tendency amongst several beneficiaries of public sector schemes tend to discontinue subscriptions to them once the liability of payment arises.

In the case of the digitisation of land records, we have already seen that unaccounted costs derailed the NLRMP. Similarly, DILRMP was

Table 3.2 Comparative Data Usage Per Capita (Exabytes/per Month)

Mobile Data Traffic	*2019*	*2020*	*Forecast 2026*	*CAGR 2020–2026*
North America	2.8	3.9	17	28%
Latin America	1.6	2.5	14	33%
Western Europe	3.1	4.4	17	25%
Central and Eastern Europe	1.5	2.2	10	27%
North East Asia	12.7	19	78	27%
China	10.2	15	59	25%
South East	3.3	5.6	32	33%
India, Nepal and Bhutan	6.7	9.6	35	24%
Middle East	1.6	2.6	18	38%
Sub-Saharan Africa	0.55	0.87	5.6	36%

Source: Ericsson Mobility Report November 2020.

given an outlay of INR 5600 crore, an amount which has already been doubled to INR 11,000 crore with the announcement of the extension of the scheme to 2024.[2] In the survey of Rajasthan's implementation, there were several prohibiting factors that were likely cost-related as well. For instance, village-level land administrators known as Patwaris did not have access to modern survey equipment or even computers.

The problem with numbers and the ready resource curse

The failings of digital interventions in land record management and agricultural extension beg the question why does the State focus so intently on the digital element of its governance and development interventions?

Digital technology allows the reporting of large aggregate numbers. As such, even if it fails to solve for any problem on the ground, it creates the illusion of something impactful. Thus, digitalised schemes enable the State to place emphasis on the volume of information going out, even though, as we have seen, these numbers mean little when considered in proper context. The nature of digital lends itself more acutely to focussing on providing access to information, rather than considering the impact of such information on the recipient. At the time of writing, the SMS total on Mkisan stood at 24,623,710,138 and the advisory count of 435,580 (Indian Ministry of Agriculture and Farmers' Welfare 2021). E-Nam boasts of the integration of 585 APMCs across 16 states in the country (Dhamecha 2019) and will soon be extended to 1000. Correspondingly, the DILRMP website boasts of computerising 6,06,531 out of 6,56,127 villages. These sets of figures seem respectable, in and of themselves. It is for these reasons that we see outspoken support for digital schemes by the State.

The widespread digitisation of government schemes and service delivery in India has given rise to a development model we dub, "dashboard-led development". Dashboards are a tool for business performance management. They are, in a nutshell, an elegant device for displaying complex and large datasets to give an indication of progress. Over the last few years, however, dashboards have gone from a way for managers to keep tabs on their workers, to dotting several websites related to digital Government schemes in the country. Mkisan, E-nam and the DILRMP, all have dashboards screaming out statistics with little to no context.

Dashboard-led development carries forward a longstanding trend of development praxis that obsesses with the quantification of outcomes. For instance, Eyben (2010) notes that in the context of international

aid, there is pressure on development practitioners to present quantitative evidence of ground-level impact to assure donors that their
money is being spent well. Results matter and often numbers are the
easiest way for donors to understand the impact of every dollar given
(Eyben 2010). On the face of things, there is nothing wrong with such
practices. Donors are entitled to ask for reports that detail what is
being done with their money. In the case of government programmes,
it allows the public sector to show citizens what public funds are
used for.

However, if institutions focus solely on quantitative footprint of
their work, it becomes problematic. Most development problems do
not lend themselves to easily quantifiable results (Eyben 2010). In the
case of the DILRMP, for instance, a major many of the issues highlighted by researchers were hidden by the patina of positive numbers
about its implementation (Shah et al. 2017). These included gaps in
laws which prevented the timely updating of cadastral maps and the
non-recording of several important parameters such as possession and
easement rights.

As the pressure to have something to show mounts, there is a tendency to abandon initiatives that do not entail concrete results (Eyben
2010). Empowerment of marginalised communities is a lofty goal, but
if it does not fit neatly into a quantifiable outcome, it is often forsaken
for something that can.

The philosophical underpinnings of the stress on outcomes and efficiency can be traced back to an idea known as substantialism (Eyben
2010). Substantialism involves considering the world around us in terms
of "quantities" or "categories" (Eyben 2010). Substantialists deploy a
mechanistic approach to problem solving (Eyben 2010). Obstacles are
"hacked", and solutions are engineered (Eyben 2010). Mechanistic approaches are useful when problems are relatively straightforward. Such
an approach works well for things like building highways, schools, or
bridges (Eyben 2010). However, this method of problem-solving runs
into trouble when faced with issues that are more complex (Eyben 2010).

To reiterate, the danger of the substantialist viewpoint dominating
development perspectives, both in the tertiary and the public sectors, is
that it encourages emphasis on the measurable, while encouraging a
neglect of things that are not (Eyben 2010). In the development sphere
and in governance, the solutions to most problems lie in unbundling a
complex matrix of closely linked factors, most of which do not lend
themselves easily to neatly reportable or measurable formats. If there is
a "need" to measure results in "numbers of kilometres of roads built or
hectares irrigated" it necessarily frustrates "the empowerment and

capacity development efforts of agencies receiving official funding"
(Eyben 2010). The disbursal of funds, then, becomes more about mea-
suring results and less about fixing the actual problem at hand. Thus, the
emphasis on quantifying outcomes begets a disregard for bolstering the
beneficiary's sense of agency in development interventions.

Digitalisation exacerbates substantialist tendencies in governance.
With the proliferation of digital, governments like donors, presume
that it is much easier to reduce problems into quantitative data points
and create "measurable" parameters. The assumption also extends to
solutions for these neatly measured problems – they are easier to
"engineer" now that there is a device in the picture. Problems that are
hard and complex can be swept under a rug of neat facts, figures and
charts. As such, it is unsurprising that neither E-Nam, Mkisan or
DILRMP make any real effort towards empowering farmers or improving the ground situation with land records in the country.

That the Indian state synonymizes the modern with the technological and further correlates the technological with progress, is illustrated by the nomenclature used for its schemes. In the case of E-nam,
for instance, in December 2019, Vasudha Mishra, the Special Secretary
in the Department of Agriculture, Cooperation & Farmers Welfare,
wrote an article in the Indian Express, a reputed national newspaper,
about how the name symbolises the virtuousness of the program, citing
how it is an "inam" (Urdu word for reward) for farmers (Mishra
2019). Similarly, the recasting of what is essentially an effort to computerise physical land records into a national and then digital modernization programme speaks to a belief in the power of technology to
rectify the human-made mistakes of the past, even though in reality it
is only serving to affix them in a different medium.

Substantialism and the general belief in the transformative power of
digitalisation is reinforced by the political economy surrounding
quantitative outcomes. Digitalisation allows the reporting of vast
numbers not only because data gathering is simplified but also because
of the proliferation of digital devices. This is especially true in a populous country like India where most numbers seem large. According
to official statistics, around 51 million farmers signed up for Mkisan
(Indian Ministry of Agriculture and Farmers' Welfare 2021). In aggregate this is greater than the population of Colombia, currently the
28th most populous country in the world ("Countries of the World by
Population" 2021). But when compared to the 2011 Census figures that
peg the number of cultivators in India at 263.1 million, it is less than
half of the country's farmer population nine years ago (PIB Delhi
2020). To anyone who is not well-versed with the ground realities of

the Indian agricultural sector, it seems like there is some value to the activity.

It is not our contention that the State means to neglect its duties to the farmers or land administration by putting out digital schemes that have little or no impact. Rather, the overemphasis on the transmission of information or computerisation of records, and consequently the volume of information sent out, reflects a governance mindset that assumes that access to information automatically converts into benefits. Success is measured by the volume of information sent out or number of records digitised because State presumes that such metrics translate into concrete development outcomes. This is apparent from the fact that the only Government studies on Mkisan, extoll the virtues of the portal by either discussing the volume of information sent out (Nagesh and Saravanan 2019) or the number of individuals registered (Anandaraja et al. 2015). There is, then, a presupposition about welfare creation without the provision of any evidence about the scheme's effects on productivity or income.

Successful development interventions should always be predicated on the ability to empower beneficiaries or address systemic issues plaguing administration. This is complicated, even with the use of technology. As it happens, technology often compounds the problem at hand because it is presumed that its involvement with an intervention will result in the automatic resolution of problems on the ground. But most development deficits such as those in agriculture in India are the result of messy dynamics between multiple stakeholders. Their resolution involves negotiating with the limitations of human cognition, the dynamics between the stakeholders in an ecosystem, and the externalities that impact the health of that ecosystem. The only means to overcome these competing considerations is good design.

If information-led development was about providing access to information to intended recipients of governance schemes, dashboard-led development is about stakeholder management. It is about having something to show, a ready data source that can be rattled off to indicate that targets are being met. However, what do you do when the metrics are off – when the thing being measured has little correlation to the problem at hand?

Blockchain-based land titling – an even more modern panacea for digital governance

Rather than course correct and implement the holistic reforms necessary to resolve problems on the ground with land records, the State now seeks to add yet another layer of technology to the melange of digital tools currently tackling the problem. A recent whitepaper released by the Ministry for Electronics and Information Technology lists the transfer of land records as a top potential application for blockchain-based governance in the country (Ministry of Electronics and Information Technology 2021). For the uninitiated, blockchains are decentralised ledgers that rely on cryptographic technologies such as digital signatures to create immutable transaction records. They rely on a peer-to-peer network and complicated consensus protocols to ensure all participants in the network are incentivised to main the ledger's integrity.

The most promising attribute of blockchains is that they effectively guard against the falsification of digital assets – whether it be money or property – because each asset is uniquely identifiable as is each transaction involving that asset.

The ostensible draw to blockchain is that it presents a technological solution for the integration of the different departments overseeing different aspects of land governance in India. There is the department of stamps and registration which registers documents pertaining to property transfer, the survey department which updates cadastral maps, and the department of revenue manages land records. At the township level, a tehsildar maintains these records and below the tehsildar is the revenue inspector. At the lowest level, is the village level administrator who is the first point of contact for anyone wishing to access or make a change to land records. There are also gram panchayats which play important roles in some states in the mutation of certain properties (Shah et al. 2017). These stakeholders tend to operate independently of one another which translates into deficiencies in record keeping.

Blockchains are presented as a solution to overcome several deficiencies in land titling, including the isolated functioning of administrative stakeholders. The general argument made in favour of the technology is one of efficiency – more operations, less time for transactions, lower cost (Thakur et al. 2019). For instance, blockchains are the autonomous execution of the tasks carried out by these different stakeholders through "smart contracts" (Thakur et al. 2019). Put simply, smart contracts are a way to program process functions so that they can be self-executing.

There are several countries that are currently testing the application of blockchains to land registries, including Ghana (Ministry of Electronics and Information Technology 2021). Interestingly, India has had its fair share of pilots in this arena as well. The states of Andhra Pradesh, Maharashtra, Telangana and the Union Territory of Chandigarh all took a stab at blockchain pilots over the last few years.

Most of these projects did not go beyond the pilot phase, with the exception of Andhra Pradesh which has seemingly scaled up its pilot and integrated it into the mainframe of its land record management system (Express News Service 2019). The hurdles were largely legal and administrative. Indeed, one of these hurdles still remains today which raises questions about the legitimacy of the Andhra Pradesh project. The Indian Information Technology Act, 2000 precludes the application of digital signatures to any transaction involving immovable property (Information Technology Act 2000). However, digital signatures form an integral part of blockchains and other decentralised ledger technology, which makes it impossible to integrate the technology into land transactions without a legal amendment.

These hurdles notwithstanding, there are more fundamental issues with the addition of yet another technological layer to land record administration. It will add another layer of complexity that officers at different levels may not be able to reckon with. The foremost among these is the fact that blockchains create ledgers that are immutable (Thakur et al. 2019). While this is a great mitigant against fraud, and could underpin a much more transparent form of governance, entry of incorrect information leaves a permanent erroneous record. As such, any immediate viability would only be for land parcels with up-to-date, accurate, and clean transaction histories. In other words, blockchains would not present a shortcut for updating legacy records and data pertaining to land – they would only provide a prospective solution.

Complexity also arises in the types of land transactions that are carried out in different states. For instance, in the Union Territory of Chandigarh, it is common and accepted practice to transfer property through General Power of Attorney (GPA). The practice was illegal. However, it became common as it enabled parties to evade registration charges on the conveyance documents (Raveendran et al. 2011). And they were able to get away with it because it is not mandatory to register power of attorney documents. Moreover, even if such documents are registered, the officer registering them is not legally required to verify their contents. The position on property sales through GPAs had been clarified by the Supreme Court in 2011. The Court established that valid transfers of property can only be carried out through property conveyance documents such as deeds of sale (Raveendran et al. 2011).

In India property transfer by GPA was so commonplace that land administrators had accepted it as customary practice. Indeed, the Government of Delhi revoked its ban on the practice in 2012, despite the standing 2011 Supreme Court order affirming its illegitimacy (Economic Times 2013).

We cite the example of Chandigarh because a state-run think tank pursued a pilot project there that involved the use of blockchains for registering land records (Kumar et al. 2020). The report issued by the think-tank acknowledged that there are legal challenges that stand in the way of such a system (Kumar et al. 2020). However, it neglects to mention the issue of property transfer by GPA. Instead, it concludes that the introduction of a blockchain-based land record management system would reduce property disputes and give people greater certainty of ownership (Kumar et al. 2020). It is uncertain, however, how such outcomes are envisioned if the messy problem of GPAs is not appropriately dealt with.

The central issue with digital agricultural extension and land record management is the neglect of emphasis on design. As such, qualitative parameters such as the kind of information sent out or the type of problem being solved for, suffer, further decreasing the utility of these interventions. The addition of blockchain technology will not solve for the myriad legal and administrative issues that hinder the rectification of the state of land records in the country. Rather, its introduction would like to create problems of its own, if not accompanied by concrete reforms in the law and the modes of land governance.

A blueprint for development: issues at the centre, technology at the margins

Design ends up being a foregone conclusion in the pursuit of digitalisation, which is seen as an end in itself, just like the notion of modernity. In the case of extension schemes, we have tried to demonstrate that the considerations around design are wasted on the mechanics of disseminating information. Similarly, in the case of land records, systemic issues related to legal amendments and administrative overhaul take a back seat to the digitisation of records.

Good design of any development scheme, digital or otherwise, uses its intended beneficiary as a starting point and works backward to consider what technology's role might be in helping along a desired development outcome. Good design is the only way to overcome a lack of interactivity in development schemes, excessive focus on quantitative reporting and to account for all relevant costs. Design can also help determine whether digitalisation is indeed the most effective way to tackle a specific problem. Digital elements in a scheme may often conflict with realities on the ground. What follows is that schemes, digital or otherwise, that are compatible with ground realities have a greater chance of success than those that militate against them.

There are certain things policymakers must consider when looking at the inclusion of technology in a governance scheme. The first is the objective. What is it we are trying to solve for? In the case of agricultural extension, the objective is to empower farmers whereas with land records the broad objective is to make land governance easier and more efficient. However, a perusal of the schemes evaluated in this chapter reveals that the State's objective has broadly been to integrate technology into an existing ecosystem. In other words, the state tends to put technology first. To recap, the state presumes that this will automatically improve the situation on the ground. However, as we have seen it generally adds complexity and makes a difficult problem even harder to solve.

Once the objective is clear, the State must ask itself whether technology has a role to play in reaching this objective. If the answer is yes, planners must consider what kind of complexity introducing a technological layer would add. This would include accessibility issues for intended beneficiaries and technical capacity limitations of the staff. The more centrally positioned a technology is in achieving a particular governance objective, the more complexity it will add, as evinced by the experience of land record "modernisation" in the country.

If technological interventions are being considered in the case of information-based scheme such as agricultural extension, planners must consider what kind of information they want to send (Aker et al. 2016). For instance, the information shared must lend itself easily to digital display formats (Aker et al. 2016). For instance, information relating to weather is easily conveyed through the digital medium, but even such basic information requires context which is often lacking on Mkisan.

Information on more practical aspects such as cropping methods, which Mkisan also provides, generally requires supplementary instruction. For instance, one Mkisan advisory suggested certain corn varieties for farmers to buy (Indian Ministry of Agriculture and Farmers' Welfare 2021). For such inputs, it is better if site visits are carried out and technology is used as a mode of communication for follow-up, if convenient. Site visits would also possibly enable officers to discern whether the farmer has ready access to a mobile phone and is capable of using it.

Next, planners must revisit older schemes and carry out a thorough analysis of exactly what went wrong. As Heeks (2001) points out, States are generally wont to carry out evaluations of their schemes. This is possibly why the Indian state has continually followed a top-down topology for agricultural extension, even though it knows it does not work.

Digitalisation can and should supplement government schemes. Specifically, digital platforms can be used to gather *relevant data* to evaluate interventions and understand their drawbacks. Technology must play an auxiliary and not a dominant role. For instance, Mkisan could have greater utility as a means of notification for farmers, sending reminders to them about what was prescribed by an expert during a field visit.

India must find a way to integrate the physical and digital in a way that both spheres complement each other. Doing so entails acknowledging the limitations of digitalisation as a solution for hard problems on the ground. The State must move technology to the margins of policy focus, to be brought in as a tool, when necessary, while bringing the problems at hand back to the centre.

Notes

1 By records we mean the Record-of-Rights (RoR) which captures the detail of the landholding such as plot size, number, name of the owner etc., the cadastral map which provides a visual depiction of the plot described in the RoR, and the register of land agreements that records property transactions.
2 https://www.livemint.com/Politics/do9qSSB3siI5VAkM4Pxe4H/Govt-nearly-doubles-budget-for-digitisation-of-land-records.html

Bibliography

Afroz, Shafi, Takhe Asha, G.R. Manjunatha, and Kanika Trivedy. "Constraints Faced by Sericultural Farmers of Murshidabad District with the Advisory Services of SMSs through MKisan Portal." *Journal of Community Mobilization and Sustainable Development* 13, no. 3 (December 2018): 554–560. https://www.researchgate.net/publication/330383754_Constraints_Faced_by_Sericultural_Farmers_of_Murshidabad_district_with_the_Advisory_Services_of_SMSs_through_mKisan_portal.

Aker, Jenny, Ishita Ghosh, and Jenna Burrell. "The Promise (and Pitfalls) of ICT for Agriculture Initiatives." *Agricultural Economics* 47, no. S1 (November 2016): 35–48. 10.1111/agec.12301.

Anandaraja, N., N. Sriram, R. Venkatachalam, and H. Philip. 2015. *Mkisan Short Message Services: A Cross Sectional Analysis in Tamil Nadu.* Directorate of Extension, Education Tamil Nadu Agricultural University. https://agritech.tnau.ac.in/pdf/15.pdf.

Annamalai, Kuttayan, and Sachin Rao. "What Works: ITC's E-Choupal and Profitable Rural Transformation." World Resources Institute, August 2003. https://wriorg.s3.amazonaws.com/s3fs-public/pdf/dd_echoupal.pdf.

Bal, Meghna. "Securing Property Rights in India through Distributed Ledger Technology." ORF. Observer Research Foundation, 5 January 2017. https://www.orfonline.org/research/securing-property-rights-india-through-distributed-ledger-technology/.

Bureau, Hindu Business Line. "Karnataka's Agri-Marketing Reforms Have Not Benefited Farmers: CAG." *The Hindu Business Line*. October 10, 2019. https://www.thehindubusinessline.com/markets/commodities/karnatakas-agri-marketing-reforms-have-not-benefited-farmers-cag/article29649290.ece.

"Countries of the World by Population." The Nations Online Project, 2021. https://www.nationsonline.org/oneworld/population-by-country.htm#MPC.

Delhi, PIB. "Agrarian Land." *Indian Press Information Bureau*. February 4, 2020. https://pib.gov.in/PressReleasePage.aspx?PRID=1601902.

Dalwai, Ashok. "Report of the Committee on Doubling Farmer's Income: Volume XI." Ministry of Agriculture and Farmers' Welfare, November 2017. http://farmer.gov.in/imagedefault/DFI/DFI%20Volume%2011.pdf.

Deshpande, R.S. "Emerging Issues in Land Policy." (Asian Development Bank, 2007), https://www.adb.org/sites/default/files/publication/30118/inrm16.pdf.

"Delhi Government Lifts Ban on Property Transaction through GPA." The Economic Times, 22 July 2013. https://economictimes.indiatimes.com/wealth/personal-finance-news/delhi-government-lifts-ban-on-property-transaction-through-gpa/articleshow/21252933.cms?from=mdr.

Dhamecha, Sheetal. "585 Mandis Integrated With E-NAM." *Krishi Jagran*, January 5, 2019. https://krishijagran.com/news/585-mandis-integrated-with-e-nam/.

ET Government, "Nirmala Sitharaman Tells States to Join the Electronic National Agriculture Market." *The Economic Times*, November 12, 2019, https://government.economictimes.indiatimes.com/news/digital-india/nirmala-sitharaman-tells-states-to-join-the-electronic-national-agriculture-market/72023436.

Express News Service, "Andhra Government to Adopt Blockchain Tech to End Land Record Tampering." The New Indian Express, 15 December 2019, https://www.newindianexpress.com/states/21ndhra-pradesh/2019/dec/15/21ndhra-government-to-adopt-blockchain-tech-to-end-land-record-tampering-2076359.html.

Eyben, Rosalind. 1 January 2010. *Relationships Matter: The Best Kept Secret of International Aid?*. CDRA Annual Digest for Practitioners of Development. A Centre for Developmental Practice. https://www.eccnetwork.net/sites/default/files/media/file/Eyben%202011.pdf.

Finger, Matthias, and Gaelle Pecoud. "From E-Government to e-Governance? Towards a Model of e-Governance." 2003. https://www.researchgate.net/publication/37423428_From_e-government_to_e-governance_Towards_a_model_of_e-governance.

Goyal, Aparajita. "Information, Direct Access to Farmers, and Rural Market Performance in Central India." *American Economic Journal: Applied Economics* 2, no. 3 (2010): 22–45. https://www.jstor.org/stable/25760218?seq=1.

Heeks, Richard. "Understanding E-Governance for Development." SSRN Scholarly Paper. Rochester, NY: Social Science Research Network, 18 February 2001. https://papers.ssrn.com/abstract=3540058.

Indian Ministry of Agriculture. "MKisan Portal: Mobile Based Services for Farmers." Government of India, 2013. https://mkisan.gov.in/images/Detailed%20Writeup%20on%20mKisan.pdf.

"India's Digital IDs for Land Could Exclude Poor Communities, Experts Warn." NDTV Gadgets 360, 2 April 2021. https://gadgets.ndtv.com/internet/news/ulpin-unique-land-parcel-identification-number-launch-rollout-march-2022-land-dispute-poor-communities-exclusion-experts-2404568.

"Information Technology Act, 2000." § 4 read with the First Schedule (n.d.), https://www.indiacode.nic.in/bitstream/123456789/13116/1/it_act_2000_updated.pdf.

Jebraj, Priscilla. "Unique ID for All Land Parcels by March 2022: Centre." The Hindu, 29 March 2021, https://www.thehindu.com/news/national/unique-id-for-all-land-parcels-by-march-2022-centre/article34184475.ece.

Kumar, Arnab et al. "Blockchain: The India Strategy." New Delhi, India: Niti Aayog, January 2020. https://static.psa.gov.in/psa-prod/psa_custom_files/Blockchain_The_India_Strategy_Part_I.pdf.

Mishra, Vasudha. "Agriculture Marketing: How e-NAM Has Become an 'Inam' for Farmers." *The Indian Express*, December 12, 2019. https://indianexpress.com/article/india/agriculture-marketing-how-e-nam-has-become-an-inam-for-farmers-6162552/.

"Mkisan." Indian Ministry of Agriculture and Farmers Welfare, 2021. https://mkisan.gov.in.

Nagesh, and Saravanan, Raj. 2019. *Impact of ICTs on Agriculture growth and Development Case Studies from Karnataka Region*. Discussion Paper 9. MANAGE-Centre for Agricultural Extension Innovations, Reforms, and Agripreneurship (CAEIRA). National Institute of Agricultural Extension Management (MANAGE). https://www.manage.gov.in/publications/discussion%20papers/MANAGE-Discussion%20Paper-9.pdf

Narasappa, Harish, et al. "Access to Justice Survey 2015–2016" (Daksh, 2016). https://dakshindia.org/wp-content/uploads/2016/05/Daksh-access-to-justice-survey.pdf.

"National Strategy on Blockchain: Towards Enabling Trusted Digital Platforms." (Ministry of Electronics and Information Technology, December 2021), https://www.meity.gov.in/writereaddata/files/National_BCT_Strategy.pdf.

Nayak, Pradeep. "Policy Shifts in Land Records Management." *Economic and Political Weekly* 48, no. 24 (15 June 2013): 71–75, http://www.jstor.org/stable/23527394.

Pandian, B.J., Sampathkumar Tharmar, and R. Chandrasekaran. "System of Rice Intensification (SRI): Packages of Technologies Sustaining the Production and Increased the Rice Yield in Tamil Nadu, India." *Irrigation and Drainage Systems Engineering* 3, no. 1 (January 2014). 10.4172/2168-9768.1000115.

R.V. Raveendran, A.K. Patnaik, and H.L. Gokhale, Suraj Lamp & Industries Pvt. Ltd. vs. State of Haryana & Anr., 2011.

Sajesh, V.K., and A. Suresh. "Public-Sector Agricultural Extension in India: A Note." *Review of Agrarian Studies* 6, no. 1 (2016). http://ras.org.in/public_sector_agricultural_extension_in_india.

Satyanarayana, A., Thiyagarajan, T. M., and Uphoff, N. "Opportunities for Water Saving with Higher Yield from the System of Rice Intensification." *Irrigation Science* 25, no. 2 (2006): 99–115. 10.1007/s00271-006-0038-8

Shah, Ajay, Anirudh Burman, Devendra Damle, Itishree Rana, and Suyash Rai. "Implementation of the Digital India Land Records Modernisation Programme in Rajasthan." National Institute of Public Finance and Policy, November 2017. https://macrofinance.nipfp.org.in/PDF/DILRMP.pdf.

Saxena, K. B. C. (2005). Towards excellence in e-governance. *International Journal of Public Sector Management*, 18, 498–513. 10.1108/09513550510616733.

Shoumitro Chatterjee and Mekhala Krishnamurthy, "Understanding and Misunderstanding E-NAM." *Seminar*, January 2019, https://www.india-seminar.com/2020/725/725_shoumitro_and_mekhala.htm.

Singh, K.M., M.S. Meena, R.K.P. Singh, Abhay Kumar, and Ujjwal Kumar. "Agricultural Technology Management Agency (ATMA): A Study of Its Impact in Pilot Districts in Bihar, India." Munich Personal RePEc Archive, March 26, 2013. https://mpra.ub.uni-muenchen.de/45549/8/MPRA_paper_45549.pdf.

Singh, Piara, D. Vijaya, N.T. Chinh, Aroon Pongkanjana, K.S. Prasad, K. Srinivas, and S.P. Wani. "Potential Productivity and Yield Gap of Selected Crops in the Rainfed Regions of India, Thailand, and Vietnam." National Resource Management Program. Andhra Pradesh, India, 2001. http://oar.icrisat.org/2511/1/994–2001.pdf.

Sulaiman, R.V. "Agricultural Extension in India: Current Status and Way Forward." Beijing, 2012. https://www.aesanetwork.org/wp-content/uploads/2018/08/sulaiman_ag_extension_india.pdf.

Sulaiman, Rasheed V., and Andy Hall. "The Fallacy of Universal Solutions in Extension: Is ATMA the New T&V?" United Nations University, September 2008. https://www.researchgate.net/publication/316736955_The_fallacy_of_universal_solutions_in_extension_Is_ATMA_the_new_TV?enrichId=rgreq-9aac7c61333dfb5df89df532862339c6-XXX&enrichSource=Y292ZXJQYWdl OzMxNjczNjk1NttBUzo0OTE5MjQ3MjM1Njg2NDBAMTQ5NDI5NTQ-wMTYwMA%3D%3D&el=1_x_2&_esc=publicationCoverPdf.

Tamil Nadu Agricultural University. "Agricultural Technology Management Agency (ATMA)." 2015. https://agritech.tnau.ac.in/atma/atma_intro.html.

Thakur, Vinay et al. *International Journal of Information Management*, 27 April 2019, 10.1016/j.ijinfomgt.2019.04.013.

Wahi, Namita. "The Fundamental Right to Property in the Indian Constitution." In *The Oxford Handbook of the Indian Constitution*. Oxford University Press, 2016.

4 Is Data Is the New Oil?

The Economist coined the term "Dutch Disease" in 1977 to describe the relationship between the discovery of large quantities of natural gas in the Netherlands in the 1960s and the consequent negative impact on the country's industrial development. Sometimes when a country discovers a valuable resource that can be exploited or traded, it may serve as a curse for other industries. The innate value of natural resources tempts governments to overvalue their currency, in a bid to earn more from trading such resources rather than industrial goods. This currency appreciation, in turn, makes industries less competitive, as their products become more expensive to buy when imported in other markets.

The Dutch Disease is also known as a "paradox of plenty" or a "resource curse", that afflicted several countries over the course of time. We argue that the massive generation of data in India, and its consequent association with natural resources like oil, has given rise to a modern variant of this paradox.

Global internet traffic grew from 100 gigabytes (GB) per day in 1992 to more than 45,000 GB per second in 2017 (UNCTAD 2019). India accounts for a large and growing share of such traffic, due to affordable mobile data. According to Ericsson, global mobile data traffic was around 456 exabytes in 2019, of which India accounted for around 75 exabytes or around 16 percent (Ericsson 2019). The country punches well above its weight in terms of mobile data consumption per capita. For instance, India's per capita consumption in 2020 was around 14.6 gigabytes compared to around 11 gigabytes in North America and Western Europe (Ericsson 2021). Indian policymakers see great potential to turn the consumption of data into a kind of modern-day gold rush. Speeches made by top public figures have compared the data to oil and even directly to gold (NDTV 2019).

DOI: 10.4324/9780429324901-4

Aside from currency appreciation, the Dutch Disease also engenders protectionism. Scholars have documented the experience of countries like Malaysia and Indonesia, which discovered natural resources in the mid-2000s, and went on to increase import barriers across industrial sectors (Sebastian et al. 2020). Similarly, the commodification of data in India's public discourse has led to protectionist thinking. The innate value ascribed to data has deprioritised policy impetus on competition and innovation. This is evident in policy discourse around themes like "data sovereignty" that are discussed subsequently. There is a growing bias towards a closed approach to digital markets, antithetical to globalisation. This will shield local companies from competition, and create dependence on policy arbitrage.

Counterintuitive as it may seem, India needs to treat data as a less valuable resource than it currently does. It should be regarded as raw material that requires dedicated innovation and creative effort, commercial investment, refinement and processing, to generate economic value. The distortionary impacts of conceptualising data as a natural resource will be hard to offset in the longer run. Conversely, appropriate policies will help the domestic digital economy to prosper. For instance, policies that incentivise risk-taking can help domestic businesses keep pace with technological change. A protectionist consensus will only foster inertia.

India has crossed 800 million broadband connections and will add millions more over the next few years. This underscores its transition to a data-rich status, from an erstwhile data-poor one. A consequence of this is that the country now sees data abundance as a sort of palliative for underdevelopment the traditional factors of production – land, labour and capital. This, in turn, embeds a rigid expectation in the policy view that data has innate value, like a natural resource. It doesn't help that this newfound abundance also creates a sense of hope, that India can catch up with its Asian rival China. The associated paradox that we attempt to highlight in this chapter is that there is little correlation between modernity as represented in the hopes of industrial progress via digital technology, and progressive changes in economic policy.

Data as a palliative for India's China envy

Many in India envy China's rapid development. For instance, a former director of a leading Indian economic institute, told the American National Public Radio (NPR) in 2010 that the Chinese Government "does things – builds roads, trains, power plants", whereas "democracy has failed us" (NPR 2010). Similarly, a former head of the China desk at the International Monetary Fund, stated that "we economists think that a benevolent dictator – a benevolent dictator with a heart in the right place – could actually do a lot of good". Nine years later, C Raja Mohan called China's development success an "awe-inspiring economic miracle" (Raja Mohan 2019). However, India has always faced complex structural barriers to achieving what China did, that are not obvious on the surface.

In 1776, Adam Smith's identified land, labour and capital as the three factors of production.[1] He contended that however complex a production process is, the price of a good is always a sum of the combination of costs of these three underlying ingredients. Gauged through this traditional economic lens, India can hope to achieve industrial production similar to more developed nations. For instance, according to Bain and Company, India's unit cost of factory labour is less than two dollars an hour, compared to about four dollars an hour for robots. This means that technically, India can outperform more developed economies with automated factory floors.

However, India was ranked 68th on the global competitiveness index in 2019, and was the worst performing among the group of BRICS (Brazil, Russia, India, China, South Africa) nations. Despite the push towards manufacturing through government initiatives like "Make in India", the share of manufacturing in the Indian economy is relatively static at less than 20 percent of GDP. Reasons for this range from poor quality infrastructure to administrative, legal and regulatory frictions to the ease of doing business.

Difficulties in the maximisation of traditional factors of production can perhaps explain some of the hope that the abundance of data now represents to the Indian polity. This is particularly the case when it aspires to compete with China. Let us first compare the two in terms of land availability per capita (Table 4.1). In 2000, India had around 38 percent of the land available per million population in China. By 2018, this figure had shrunk to around 32 percent on account of

Table 4.1 Square Kilometres of Land per Million Population

Year	India	China
2000	2813.99	7435.36
2010	2408.84	7018.15
2018	2198.10	6740.87

Source: World Development Indicators.

varying rates of population growth. There are accounts of growing per capita land scarcity within official documents too. According to the 2016–17 Economic Survey of India, India's land-to-population ratio will decline fourfold by 2050, relative to 1960. This will make it among the most land-scarce countries in the world.

A low per capita availability of land manifests itself in higher fixed costs for industry. For instance, an acre of (agricultural) land in India can cost anywhere between INR 500,000 to 6,000,000, whereas, it costs around INR 225,000 in Germany (Singh 2016). This is on the basis of a market exchange rate. Land costs in India seem even steeper when compared on the basis of purchasing power parity – or the exchange rate at which the currency of one country would have to be converted into that of another, to buy an equivalent amount of goods and services in each country.

Second, both India and China have vast reservoirs of labour. Together, they account for about 40 percent of the working-age population of the world. However, all workers are not equal. The productivity of labour depends on investments made by countries in areas like education and healthcare – which can also be termed as investments in "human capital". There is a large body of contemporary economic literature that suggests that human capital is one of the most important determinants of economic growth.[2] These deviate from the classical economic theory which centres on Smith's factors of production and regards labour productivity as an external or exogenous factor.

There are large divergences in the quality of labour or human capital between China and India. Education is a good proxy for highlighting these differences, because of its strong correlation with human development outcomes. For instance, countries at the top of the Human Development Index are much more likely to school citizens for longer than countries at the bottom. The differences in educational outcomes across India and China are stark. Less than one percent of Indians in the 15–24 age group completed secondary education compared to close to 73 percent of Chinese (Table 4.2). There are also order of magnitude differences in improvement of these outcomes between 1970 and 2010. These differences are not visible in the average years of schooling, since India has done well on enrolment and citizens who now have a constitutional right to education. However, there is little improvement in completion of secondary education, an example of the persistence of rather poor quality of basic public services.

Third, despite liberalisation and digitalisation of finance, India lags much behind China in development of capital markets. The International Monetary Fund's Financial Development Index is a

Table 4.2 Educational Attainments of Population Aged 15–24 Years in China and India

Country and Year	% of 15–24 Population in Group of 15 Years and Above	No Schooling	Highest Level Attained = Primary		Highest Level Attained = Secondary		Highest Level Attained = Tertiary		Average Years of Schooling
			Total	Completed	Total	Completed	Total	Completed	
			(% of population aged 15–24 years)						
People's Republic of China									
1970	31.6	16.3	44.8	30.5	38.4	7.4	0.5	0.1	5.5
2010	20.0	0.1	3.5	2.3	75.9	72.8	20.5	6.1	10.9
India									
1970	30.4	52.3	37	24.3	9.6	0.2	1.1	0.4	2.5
2010	27.1	7.1	25.6	25.6	61.8	0.9	5.5	1.8	7.1

Source: Lee, J.W and Francisco, R. "Human Capital Accumulation in Emerging Asia 1970 – 2030". *Japan and the World Economy* 24, no. 2 (2012): 76–86.

Table 4.3 Financial Development Index, 2017

	Overall Index Value	Financial Market Depth	Financial Market Access	Financial Market Efficiency
China	0.64	0.70	0.24	1.00
India	0.42	0.59	0.20	0.54

Source: International Monetary Fund.

relative ranking of countries on their financial institutions and markets. Table 4.3 below highlights some useful index parameters. The measure of financial market depth compiles data such as stock market capitalisation, total debt security and stocks traded, as a proportion of GDP; whereas the measure of financial market access looks at market capitalisation outside of top 10 largest companies and total number of issuers of dept per 100,000 adults. While India is placed close to China in terms of access metrics, it is far behind in terms of market depth. A lack of financial market depth has an impact on more composite metrics such as financial market efficiency which compiles data on stocks traded relative to market capitalisation, as well as on the overall value of the relative index.

India's underperformance in unlocking capital as a factor of production is also acknowledged in official documents. For instance, the 2019–20 Economic Survey highlighted that "*India's banks are disproportionately small, compared to the size of its economy. In 2019, when Indian economy is the fifth largest in the world, our highest ranked bank—State Bank of India— is a lowly 55th in the world and is the only bank to be ranked in the Global top 100*". In contrast, China has some of the largest banks in the world, including four among the top 10.[3]

The underlying deficits in seemingly abundant factors of production is a stark challenge for India because they are each required for industrialisation to compete with China. Conventional economic theory suggests three stages of development that are characterised by dependence on agriculture, manufacturing and services. The third stage is also associated with de-industrialisation as resource allocation shifts to services. That is, industrialisation peaks as it enters the third stage of development and then begins to decline. However, economists such as Dani Rodrik have observed that developing countries seem to be achieving peak industrialisation at lower levels of per capita prosperity than advanced counterparts (Rodrik 2015). Advanced countries like Britain and Sweden saw industrialisation peak at around USD

14,000 per capita (in 1990 dollars), whereas India appears to have reached its peak at around USD 700 (Rodrik 2015).

Achieving peak industrialisation at low levels of economic development and prosperity is problematic primarily because India sees over a million people enter its workforce every month. Only industrial manufacturing activities are labour intensive enough to accommodate such a large demographic bulge. But, as the 2014–15 Economic Survey notes, "the sobering fact is that India seems to be de-industrialising too" and that "to call the Indian phenomenon de-industrialisation is to dignify the Indian experience, which is more aptly referred to as premature non-industrialisation because India never industrialised sufficiently in the first place". The Survey also references Singaporean leader Lee Kuan Yew's address at the Jawaharlal Memorial Lecture in New Delhi in 2005, where he pointed out that "since the industrial revolution, no country has become a major economy without becoming an industrial power". Such circumspection is rare in official documents.

Clearly, India's policy elite is familiar with the scarcities of cheap land, quality labour and long-term capital. The Indian Parliament has also seen several discussions on the various indicators of such scarcities. In March 2020, Member of Parliament (MP) Thalikkottai Rajuthevar Baalu stated in Parliament that the number of people without private property is "rising and the ones without property joined the ranks of the worst ones in extreme poverty and the task of poverty alleviation has become even more difficult". In 2019, MP Ravneet Singh spoke in Parliament to highlight that "agricultural labour productivity in India is less than one-third of that of China". Similarly, in 2020, MP Gowdar Mallikarjunappa Siddeshwara noted that the MSME sector in the country is "severely under-capitalised and is in need of capital infusion for its survival". While these scarcities are in constant focus, factor market reforms remain an unfinished project for the Indian State. In fact, even in the immediate fallout of the COVID-19 pandemic, the Prime Minister renewed focus on these factors of production (Modi 2020).

Despite the fact that India has been unable to leverage any factor in which it has a surplus in the past, decision makers feel more self-assured about their ability to leverage data as a new factor of production. This reflects not just in the Government's ambitious target to achieve a trillion-dollar digital economy by 2024–25, but also in related statements in the Parliament. In 2017, the junior minister for IT highlighted with confidence that "the digital economy is already contributing to the growth of industrial sector" and that "the operational efficiency and benefits to all sectors of the economy, including the industrial sector, will further increase with greater use of digital technologies" (Chaudhary 2017). Data is also at the core of the most valuable multinational corporations such as Amazon, Google and Facebook. Consequently, policymakers are compelled to consider it as an additional factor of production. They are not alone. In 2012, a report by the Economist Intelligence Unit, highlighted that nine times out of ten, business leaders in developed markets believed that data "is now the fourth factor of production" and is as fundamental to their businesses as land, labour and capital (Jones 2012). However, the more India thinks of data as a palliative for the underlying deficits in traditional factors of production, the more it is likely to become prey to a modern-day variant of the Dutch Disease.

India's optimism around data partly stems from the successful privatisation of telecom, which led to some of the lowest voice and data tariffs in the world. The country's data consumption has expanded exponentially over the past decade, and unlike traditional factors of production, has kept pace with China's (Table 4.4). For instance, in 2013, India accounted for a two percent share of global data consumption, and China accounted for five percent. In 2019, these shares grew to 19 percent and 31 percent respectively. The combined digital might of the two countries also manifested in 2019, since they accounted for half of global data consumption. These are all seemingly legitimate reasons for India to celebrate the dazzle of the digital!

India's early celebrations are encouraged in international forums, where the country is seen as a possible counterweight to China in the long-term. According to the World Economic Forum, which popularised the concept of the Fourth Industrial Revolution, the abundance of data "can help India leapfrog traditional phases of development and

Table 4.4 Mobile Data Traffic (Exabyte/Months)

Year	World	India	China
2013	2	0.049	0.096
2014	3.2	0.044	
2015	5.3	0.164	0.385
2016	8.8	0.926	1
2017	15	2.238	2.7
2018	27	4.867	8.6
2019	38	7.176	11.8

Source: Cisco, TRAI, GSMA, Ericsson.

accelerate its transition to a developed nation" (*Economic Times* 2018). The appeal of leapfrogging is attractive, and India's polity has endorsed this view. Prof Jayshree Sengupta documented this view succinctly, "the Fourth Industrial Revolution is possible in India, according to official sources and industrialists like Mukesh Ambani, because of the huge increase in data usage and the Digital India campaign" (Sengupta 2017). The World Economic Forum also seems to salivate at the prospects of deploying algorithms on "behavioural data of 1.25 billion Indians across the 19,500 classified languages and dialects, nine recognised religions, and a variety of races, castes and sects". Sengupta pointed out that industrial revolutions are complex processes because they involve transformations of "entire systems of production, management and governance". These aspects are hard to gloss over, but the promise of a data-driven economy offers a new paradigm for India's future industrialisation.

Data as the new oil

We trace how data became associated with oil in economic discourse since this view of data shapes India's technology policy approach and strategic calculus vis à vis China. Comparisons between data and oil trace back to corporate marketing forums in the mid-2000s. Clive Humby, a self-styled data scientist, first used the phrase "data is the new oil" at a marketer's summit at the Kellogg School (Palmer 2006). The context of this assertion was that data, like oil, requires refinement and processing to produce value. Humby runs a consulting firm that works with a range of B2C companies and analyses customer data. Therefore, his focus on refinement of data drew from his own capabilities to transform raw customer data into what he calls "*business insights*". Humby's data expertise went into products like the "Tesco Clubcard", which is a popular loyalty card of the British supermarket chain Tesco.

In India, the use of the term "*data is the new oil*", often strays far from its original context. In 2019, the Indian Prime Minister, Narendra Modi, made a strong pitch to US investors by calling data the new oil, and adding that "*data is the new gold*" (NDTV 2019). He explained this by stating that mobile data tariff in India was "cheapest in the world". Modi's comparison of data to oil, and then to gold, suggests that data has innate or intrinsic value. This deviates from Humby's original take that the value of data only emerges once it is processed. At the same time, Modi also seems to indicate an opportunity for some sort of price arbitrage on Indian data by highlighting its affordability and abundance. This is similar to the commodification of cheap IT labour that became the hallmark of the Indian IT story since the 1980s.

Dependence on a cheap IT workforce eventually precipitated structural challenges in India. In 2019, an Executive Vice President at Cognizant, an American IT multinational with a large India presence, declared that Indian IT is on the "cusp of change" (*Economic Times* 2019). The change he referred to was from an outmoded labour arbitrage model fuelled by a cheap IT workforce, to an intellectual arbitrage model characterised by innovative IT products and services. Several CXOs validated this view. In the same year, Oracle India's Managing Director noted that India's IT services "is witnessing one of its most turbulent times" due to the advent of new technology, and the need for reskilling talent (Kumar 2019).

That rivals like Philippines and Vietnam trounce India in business process outsourcing (BPO) in global markets, the low cost and low skill backbone of India's IT exports is clear to government officials who are privy to hard evidence. For instance, the transcript of a Ministry of Commerce and Industry meeting held in December 2017, suggests that officials were alarmed at the fact that IT services were "*running out of steam*" as competition became tougher in low-cost segments (Patel 2018). An official highlighted that "*technology is making many BPOs redundant*". It is paradoxical that India sees innate value in cheap data, even as cheap IT labour is no longer a prized asset. The country is likely to underinvest in innovation linked to data processing, if the abundance of raw data is viewed from narrow, opportunistic lens.

In 2019, the United Nations Conference on Trade and Development warned that "countries with limited capabilities to turn digital data into digital intelligence and business opportunities are at a clear disadvantage when it comes to value creation" (UNCTAD 2019). This warning should be taken seriously as UNCTAD is a friend to developing countries. It was established in 1964, to represent the interests of the Group of 77, a coalition of developing countries including India, in a world dominated by developed countries. The UNCTAD also highlights that "to prevent increased dependence in the data-driven global economy, national development strategies should seek to promote digital upgrading (value addition) in data value chains, and to enhance domestic capacities to refine the data". India needs a cogent and responsive digital strategy that helps achieve this. Instead, the most vocal and visible public and private sector rhetoric has encouraged protectionist policy.

Calls for protectionism can be seen in policy discourse on sovereign ownership over data. "Data sovereignty" became a buzzword in the advanced world after Edward Snowden revealed that he was the source of leaks from the US Government's surveillance programme called Prism. The aftershocks of these leaks manifested themselves in policymakers in advanced countries expressing the "business risk" associated with offshore storage of data. The Snowden leaks showed that countries can access and monitor business data without user or business consent. Consequently, data sovereignty emerged as a top ICT concern in a survey of a thousand decision makers in the US, UK, France, Germany, and Hong Kong, conducted in 2014 (*The Guardian* 2014).

While advanced countries look at the need for data sovereignty in the context of surveillance, India looks at it as an economic arbitrage opportunity. In 2019, the country's richest man, Mukesh Ambani, stated that "*India's data must be controlled and owned by Indian people — and not by corporates, especially global corporations. For India to succeed in this data-driven revolution, we will have to migrate the control and ownership of Indian data back to India — in other words, Indian wealth back to every Indian*" (Johari 2019). This concept of data sovereignty, like the notion of data as oil, deviates from the context in which the term was originally used in other parts of the world. This is not just borne out by Mr. Ambani's assertions. The notion of data sovereignty also found mentioned in the first public draft of India's eCommerce policy. The draft stated that "the increasing importance of data warrants treating it at par with other resources on which a country would have sovereign right" (Ministry of Commerce and Industry 2019). By definition, an eCommerce policy sets the governance context for commercial transactions in the digital economy. In this sense, the treatment of data as a sovereign resource indicates the impulse towards localisation of data-driven businesses.

Like the mischaracterisation of data as oil, many in India frame the data sovereignty debate incorrectly. The notion of data sovereignty as an economic strategy contradicts local economic realities. For instance, Indian IT companies, which are the backbone of the digital economy, derive most of their revenues from global markets. In fact, IT services account for a 45 percent share in India's total service sector exports. Conversely, domestic market revenues in export-oriented IT industries such as BPOs, are a fraction of export revenues (Table 4.5). India is the BPO hub of the world and if it hopes to sustain its IT services strength, it will need to mature into an IT innovation hub. This would mean that Indian digital services would need global market

Table 4.5 Revenue Trends in ITeS-BPO Industry (INR Crores)

Year	2012–13	2013–14	2014–15	2015–16	2016–17(E)	CAGR% (2012–17)
Domestic Revenues	17,500	19,593.8	21,490	23,364	26,800	12.53
Export Revenues	99,570.1	1,23,423.9	1,37,573.1	1,59,743.1	1,74,387.2	10.34

Source: Available/Downloaded at/from Ministry of Electronics and Information Technology, Government of India, sourced from NASSCOM.
Note: 'E'-Estimate.

access – or that the world would need to outsource its IT innovation to India. In either case, India would need to handle data of foreign citizens. However, an economic connotation to data sovereignty forecloses this option, because global trade and commerce is built on the foundations of reciprocal behaviour. This realisation perhaps led to more diplomatic tiptoeing by a former executive head of the telecommunications ministry, who told the New York Times in 2019 that "we don't want to build walls" while giving the political caveat that "at the same time we explicitly recognize that data is a strategic asset" (Sherman 2019).

The bipolar digital economy

Most of the largest digital companies in the world are based either in the US or China. This is a cause of worry for Indian policymakers if the country is to catch up with China in the Fourth Industrial Revolution. The promise of data is linked to the country's ability to harness digital technology to create global companies. At the same time, the comparison of data with oil can hurt India's prospects to address deep-rooted challenges that erode economic competitiveness. For instance, it is evident that to catch up with China in the digital sphere, Indian companies will need to generate similar economic value as Chinese ones. However, new digital markets offer no silver bullet – they do not give India a competitive edge in terms of outsized value generation compared to global peers. Digital markets can reinforce the same inequities between countries as traditional markets, In the absence of structural changes to the economy through the development of traditional factors of production,

India's top 10 digital companies are valued at about 10 percent of Chinese equivalents (Table 4.6). The top 10 Chinese companies are valued at around 26 percent of the equivalent companies in the US.

A standard argument against like-for-like comparisons as made here, between digital markets in different economic settings, is the difference in levels of prosperity of competing economies. The most prominent of such differences is per capita income, which is also a determinant of the appetite to buy goods and services in any economy. Therefore, the per capita income in a market is a close proxy for consumption. However, even when considering per capita incomes, the value generated by large Indian digital companies is asymmetrically low compared to Chinese counterparts. India's per capita income is about 44 percent of China's, and China's per capita income is about 24 percent of America's. That is, Indian per capita income is much

Table 4.6 Valuations of Top 10 Digital Economy Companies in US, China and India

Indian Company	Valuation or M.Cap ($bn)	Chinese Company	Valuation or M.Cap ($bn)	USA Company	Valuation or M.Cap ($bn)
Reliance Jio	65.150	Alibaba Group Holding Limited	578.850	Apple Inc.	1397.000
Bharti Airtel	41.370	Tencent	525.859	Microsoft Corp.	1387.000
Flipkart	20.000	Meituan Dianping	119.010	Amazon Inc.	1227.000
One97 Communications	17.000	JD.com	86.006	Alphabet, Inc.(Google)	946.960
Ola Inc.	10.000	Pinduoduo	79.367	Facebook, Inc.	644.723
Snapdeal	7.000	Bytedance	75.000	Netflix, Inc.	182.224
Mindtree	2.020	Didi Chuxing	56.000	Paypal	181.497
BillDesk	1.800	NetEase Inc.	52.780	Salesforce.com, Inc.	154.105
Pine Labs	1.600	Lufax	39.400	Booking	71.328
Policy Bazaar	1.500	Baidu	37.589	Uber Technologies, Inc.	63.169

Author's compilation based on open source financial data, as on 5 June 2020 11 AM IST. Selection based on "Digital Economy Report", UNCTAD, 2019.

closer to China's than Chinese per capita income is to America's. However, the top 10 Chinese companies are valued at over a fifth of similarly ranked American companies. The valuation of the top 10 Chinese companies is about the same fraction as the per capita income, relative to America. When Indian companies are compared in similar terms, they are seen to significantly underperform. That is, they should be valued at around half of what Chinese companies are valued at, given that Indian per capita income is at about half of China's. However, as previously noted, they are actually valued at around a tenth of China's top 10 digital companies. Three fundamental challenges or differences underpin this asymmetry.

The first challenge lies in the value that India generates in traditional markets, where it produces more output than the US and China. The movie market is a good proxy for this, as cinema is now available on digital platforms. In 2016, India produced nearly 500 more feature films than the US and China combined. Yet, the total revenues from this prolific output was less than 20 percent of what China alone was able to generate in its film industry (Figure 4.1). This high volume and low-value paradigm exemplify India's inability to monetise and commercialise its economic output.

Digital platforms don't necessarily help offset underperformance in value creation in traditional markets. India has close to 50 video streaming platforms – more than most developing countries in the world – most of which stream local movies and shows digitally. Yet, average film revenues are nowhere close to other markets (Figure 4.1). The monetisation gap for Indian films partly stems from a lack of

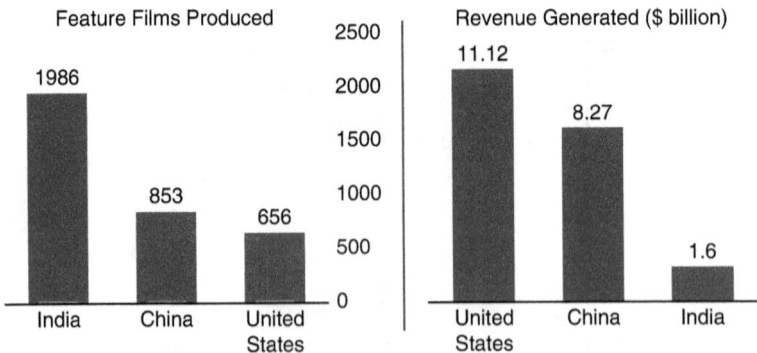

Figure 4.1 Film Output (2016) versus Revenue Generation (2017).
Source: UNESCO.

traditional infrastructure – such as cinema halls. Screen density in India is the lowest in the world. The country has around 10 screens per million people as against 124 in the US and 90 in China (*Times of India* 2017). The other part of the challenge is that the average value of digital transactions is low.

Wide gaps in digital transaction volumes and values help identify a second deviation from Chinese markets. That is, India still lacks an efficient digital market which can aid in the monetisation and commercialisation of goods and services online. Conversely, China's hyper-efficient digital transaction ecosystem is a key reason for its relative outperformance (Figure 4.2). In fact, China even outperforms the US in terms of the volume of digital transactions. This is perhaps why its digital economy has yielded economic value commensurate with its per capita income relative to the US.

A third area where China has a head start over India is in the generation of intellectual property (IP). The protection provided to innovation through IP makes it the bedrock of value generation in the digital economy, where the marginal cost of production is low. That is, the corollary to the ease with which products and services are produced and distributed in the digital ecosystem is that market competition is higher and innovation is a necessary differentiator. For instance, unlike physical markets where certain producers can have a geographic monopoly because of their distribution capacity and reach,

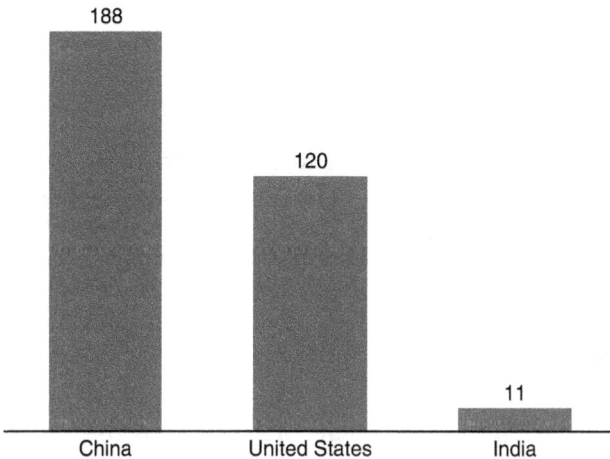

Figure 4.2 Volume of Digital Transactions in 2018 ($ Billion).

Source: Bank for International Settlements.

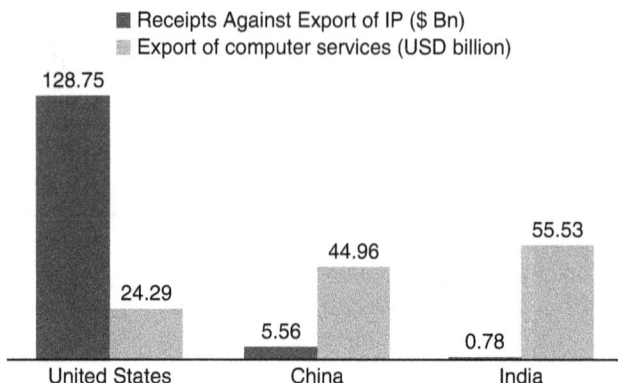

■ Receipts Against Export of IP ($ Bn)
▨ Export of computer services (USD billion)

Figure 4.3 IP and Computer Services Exports (2018).
Source: WTO.

digital markets make for a more level playing field where this is not a constraint. Royalties received against IP exports are therefore a useful proxy to measure the relative long-term dynamics of innovation in digital markets (Figure 4.3). It is evident that India accounts for a negligible fraction of the size of IP exports of the US and about 14 percent of China's.

Why is it that despite low focus on IP so many in India are confident of harnessing digitalisation towards a new form of industrialisation? Perhaps one reason is India's relative outperformance of US and China in computer service exports, which is also shown in the figure. However, computer service exports from India are largely not IP driven as they tend to involve services to solve software problems, provide backend product development and testing support, resolve consumer grievances and process data for overseas clients. Moreover, as discussed earlier, the competitiveness of Indian computer services exports is eroding, and the country will need to pivot to an innovation-led exports framework, with IP at its core. This will require a nuanced understanding of the drivers of competitiveness in digital markets.

Platforms and Innovation – Twin pillars of digital markets

A combination of asymmetrically low-value generation from production of goods and services, lack of depth in digital financial markets, and inadequate focus on innovation converge to hamper India's ability to compete with China (and the US) in digital markets. They also impair the country's capability to create large digital companies – which capture the global market. In 2020, for every large digital firm in India, characterised as firms valued at over a billion dollars by private or public markets, there were five Chinese and 13 American counterparts. It's important to underscore that large digital firms like Facebook or Google tend to platformise. That is, they perform multiple functions and provide multi-utility interfaces to their users. Such digital platforms unlock both value and volumes in digital markets through enabling transactions for goods and services. They determine how a large share of the global population communicate, transact, search for information, search for services, buy consumer products, find new jobs, store data, distribute and market products and so on.

In 2019, seven of the eight largest digital platforms, in terms of market capitalisation or valuation, were based in China or the US. This is because both countries are at the frontiers of technology innovation. According to the UN, they account for 75 percent of all patents related to blockchain, 50 percent of global spending on Internet of Things (IoT), and more than 75 percent of the global market for public cloud computing. It's no surprise then, that they also account for 90 percent of the market capitalisation of the 70 largest digital platforms, that tend to combine the utility of such technologies (UNCTAD 2019). Indian policymakers seem to be missing the woods for the trees by pinning all hopes on data collection, without focus on innovation to leverage such data through multi-utility platforms. It's true that user data sits at the core of digital platforms. It allows them to use the data collected for one function, for several others, and therefore enables economies of scale. For instance, a platform such as Google can use its vast archives of data based on its search function, for other services such as targeted digital advertising that it may offer to other users looking to use the platform for marketing a product or service. However, access to user data is just part of the picture.

The ability of American and Chinese companies to create globally dominant, multi-functional platforms, stems from the ability of companies in these countries to take risks that lead to innovation. Chinese and American digital economies are both characterised by "a willingness to push things forward" and a "high level of risk tolerance", respectively (Hermes 2020). These characteristics are virtually interchangeable, and suggest a consistent attitude towards risk-taking in the digital economy, but are underpinned by different drivers.

In the US, entrepreneurship and failure are seen as two sides of the same coin. Therefore, failure is seen as valuable experience, and the ability to push legal boundaries through new business models is closely tied to a celebration of the spirit of innovation – in other words "Americans did not wait until they knew something was going to be legal; they just made sure that it was not already illegal" (Hermes 2020). Conversely, failure is associated with public disapproval in China and therefore not encouraged. However, the rivalry with the US in the digital space has pushed the Chinese digital mindset to focus on succeeding at all costs. Competition with the US for "global technological supremacy is an immense incentive and provides the motivation for China to catch up with American platforms and push them off the throne" (Hermes 2020).

Cross-cultural differences in risk-taking are undoubtedly also shaped by incentives like social security. Risk aversion can be triggered by both an absence of meaningful social security, or an over-dependence on it. For instance, the idea of a welfare state has been blamed for reducing incentives for work (Sinn 1995). Simultaneously, social security can also increase risk-taking appetite of citizens, since the state has a stake in the outcomes of people's business decisions. Conversely, risk-taking capabilities cease to exist with a complete absence of a welfare net. This is because individuals are unlikely to take risky business decisions in environments of uncertainty – as became clear during the COVID-19 pandemic in India when entrepreneurship came to a veritable halt (Peermohamed 2020).

According to the economist Rathin Roy, India is unlikely to be able to put in place a universal basic income – a universal safety net of sorts – because of its fiscal constraints. He argues that for any semblance of such a safety net to exist, India would either have to raise taxes, increase borrowings, or spend less on existing commitments (Mishra 2019). An absence of meaningful social security, low per capita wealth, or a state-supported competitive ethos, means that India will have to come up with other policy measures to encourage risk-taking that can drive innovation. That is, encouraging entrepreneurs and businesses to take risks, should be an explicit goal of policymaking for digital. However, the conceptualisation of data as oil or data as a sovereign resource is antithetical to this cause.

Unintended consequences on policy design

Digital businesses prompted fundamental shifts in the Indian and global economy over the last decade. This led to the emergence of new modes of communication and information-sharing, new business models and new sources of job growth. These changes inevitably led to fresh policy formulations and regulatory concerns. A key characteristic of digital businesses is the ability to innovate new products and services, which allows them to shift seamlessly between regulated and unregulated markets. Therefore, the state is always playing catch up and India is no different.

Traditional legal-regulatory frameworks, based on licences and controls, suffer from lack of agility and leeway to accommodate the pace of technological development in the digital economy. Governments around the world are exploring ways to address this problem. They are debating the merits of wider goals-based or principles-based regulation versus narrower rules-based regulation. The former requires thoughtful calibration and state capacity for enforcement but is more resilient to changes in technology and business models. The latter allows for easier enforcement but is rigid and stymies innovation. However, a lack of appropriate guiding principles can also result in unintended consequences. For instance, the conceptual frame of "data as oil" leads to an incumbency bias, just as it does in extractive commodity markets.

Extractive markets such as oil and gas, or metals mining, are usually dominated by firms that have a geographical monopoly. These monopolies are sometimes state-determined, through the limited issuance of exclusive licences or permits for exploration and extraction based on regions. They are also natural monopolies, because of the large costs involved in extractive activities, including corrupt practises to seek political patronage. For instance, according to the OECD, one in five cases of transnational bribery occurs in the extractive sector (OECD 2016).

An incumbency bias in policymaking can easily translate to an uneven playing field for new businesses and entrepreneurs. In the digital economy, incumbents tend to be companies associated with running telecom infrastructure, which is a costly activity like commodity extraction, and survival in this market requires deft political management. Unsurprisingly, Indian telecom service providers (TSPs) have often asked for more regulation of the digital economy – and the regulator has responded positively. For instance, in response to a 2015 consultation by the Telecom Regulatory Authority of India (TRAI), on the governance of Over the Top (OTT) services or digital applications, telecom companies asked for a level playing field through more regulation and oversight over new markets.

Airtel, one of India's largest private operators, even proposed the concept of "Net Equality", in a submission to the TRAI in 2015, to argue for more regulation of new services rather than deregulation of the old. It stated that "the primary concern of TSPs is the absence of any regulatory parity". It added, that the "need of the hour, therefore, is compelling action prompted by an urgent need for Net Equality". Similarly, the state-owned operator, Mahanagar Telephone Nigam Limited, submitted that digital applications should "not be allowed to offer voice services in India till regulatory level playing field is established both in respect of network related issues and commercial aspects", to TRAI. If this recommendation was taken seriously then, services like Skype and Zoom, which played an invaluable role in helping people work from home during the 2020 COVID pandemic, would need licences from the Department of Telecommunications (DoT) and would be regulated by TRAI. This may have reduced their incentives to roll out business and technology innovations in India, given that they operated without licensor's supervision and oversight in most parts of the world!

The incumbent's impulse to maintain a foothold over new markets through more regulation is closely tied to the conceptualisation of data as oil. Conversely, if data was not associated with some innate value, perhaps policymakers would be biased towards fostering innovation. For telecom operators, this would translate to less regulation of their own ecosystem rather than more regulation of competitors. For instance, Indian telecom companies are subject to a wide range of taxes and levies, that far outweigh any such obligations of digital companies. These could easily be reduced or repurposed to give telecom companies the opportunity to reinvest profits into innovation. They could be given tax sops for investments in research and development instead of paying the exchequer.

Second, comparisons of data with oil can also prompt an impulse to regulate prices in digital markets, much like the price of refined oil when it is sold to end-users. This would spell disaster for the Indian digital economy. There is already evidence of regressive impulses to regulate prices in the digital economy.

The Indian TV market, which in some ways is the precursor to online distribution of video, is also governed by the telecom regulator, the TRAI. The TRAI, in turn, has imposed economic controls on the TV market since 2004, using a combination of price ceilings on channels, and restrictions on the manner in which broadcasters can offer their channels to consumers in bundles or bouquets. For instance, as per the latest TRAI regulation in force as of July 2020, channels that are part of bouquets cannot be priced at more than Rs. 19, and High Definition channels cannot be bundled with standard definition ones. Similarly, free-to-air channels cannot be bundled with pay channels in the same bouquets. Imagine if digital applications are regulated like TV. It would mean that platforms that offer free content would not be able to simultaneously offer subscription content. It would also mean that while subscription bills may decrease, the quality of content may begin to mirror that on TV. This is because there would be no economic logic left for producing differentiated content (Mittal 2020).

The TRAI is often petitioned by TV distributors to apply this pre-scriptive regulatory framework in the digital applications sphere, so that new competitors are stopped in their tracks. These distributors feel threatened by the rise of Over the Top (OTT) streaming services like Netflix. For instance, in response to a 2019 consultation paper issued by TRAI, the Cable Operators Welfare Federation of India, an industry association for TV distribution platform operators (DPOs), stated that digital applications "are mushrooming in absence of any registration and regulatory guidelines to follow, while are providing services that can be regarded as same/similar to services offered by TSPs and DPOs". Similarly, Dish TV, one of India's largest DPOs, asked for more fetters on online creative expression. It submitted that

> it is pertinent to point out that unlike other broadcasting services, no content regulation is applicable on OTT services and as a result thereof the content being delivered through such services is totally unregulated and there have been lot of complaints that in certain OTT services the content being shown is not compliant with the statutory advertising and programming code stipulated by the government.

Economic regulation can also become a means for nationalisation of private data businesses in the future, an idea that first began to manifest in policy thinking around source codes and algorithms. The Draft eCommerce policy, issued by the Ministry of Commerce and Industry in 2019, made references to the prospects of regulation of source codes and other commercially sensitive data (Ministry of Commerce and Industry 2019). It stated that "policy space must be retained to seek disclosure of source code for facilitating transfer of technology and development of applications for local needs as well as for security". This impulse to retain "policy space" or the state's dis-cretion to intervene in digital markets, is the reason for preponderance on data sovereignty as a guiding principle. In the case of the Draft Policy, the principle was employed to suggest mandatory disclosure of source codes, which is akin to expropriating private property of businesses, or a kind of data nationalisation. Source codes sit at the heart of data businesses. They shape the functions of most digital applications. They are also protected under IP regimes across the world and in India.

The threat of nationalisation is an obvious disincentive to risk-taking and innovation in the digital economy. And it keeps coming back in various forms. For instance, in 2020, the Ministry of Electronics and Information Technology (MeiTY) constituted a committee to look into the formation of a governance framework for "non-personal data" (NPD). NPD is all data outside the scope of personal data that can identify an individual, such as names, addresses, and so on. Like the Draft eCommerce Policy, this NPD committee also subscribed to the notion of data sovereignty for inspiration. It suggested that data sovereignty is the basis for establishing legal rights over NPD, and therefore "the laws, regulations and rules of the Indian State apply to all the data collected in/from India or by Indian entities" (MeiTY 2020).

The correlation between data sovereignty and weakening of property rights is fairly straightforward. The NPD committee spells it out. It dilutes the notion of private ownership by stating that the term only "holds full meaning only in terms of physical assets" (MeiTY 2020). This is the first body blow to any commercial interest in investing in a data business, where ownership rests on the uncertain principle of data sovereignty. It additionally adds that for data, ownership is "relatively loosely employed to mean a set of primary economic and other statutory rights". Therefore, it derives a notion of "beneficial ownership/ interest", to ensure that a "community's interests are safeguarded regarding non-personal data over which there is an expectation of benefits being accrued to itself" (MeiTY 2020). This is the second body blow since the Committee essentially creates an avenue for any group of people to claim "interest" in privately held datasets. One public commentator remarked that "the message from the committee to entrepreneurs and businesspersons in India is: the Indian state or a data trustee (which could be a government department) has the rights over the data your company collects or creates" (Pahwa 2020).

Linking the regulation of data businesses to the notion of data sovereignty also leads to economic protectionism – or the discrimination between domestic and foreign enterprises. This is visible in both the Draft eCommerce Policy and the recommendations of the NPD Committee.

The Draft eCommerce Policy states that "domestic alternatives to foreign-based clouds and email facilities will be promoted" (Ministry of Commerce and Industry 2–2019). It also states that "Policy space to grant preferential treatment of digital products created within India must also be retained". These statements are akin to the post-independence industrial policy approach, wherein India sought to "pick winners" and provide state support for domestic industrial ventures to compete with global entities.

The public intellectual, Swaminathan Aiyar, describes the impulse for economic protection well. According to him, notions like "economic independence" and "import substitution", in the quest to find domestic alternatives to foreign goods and services, were premised on a theory that "infant-industry protection would ultimately make India a great, competitive industrial power" (Aiyar 2018). He adds that

> industrial licenses were used to tightly regulate all production, and imports competing with newly licensed items were banned or taxed at high rates often exceeding 100 percent. This approach failed to create world champions and instead created uncompetitive high-cost industries that harmed consumers and investors alike. (Aiyar 2018)

The combination of an incumbency bias that results in calls for licencing and permissions regimes in the digital economy, and the notion of state support to select industries based on the whims and fancies of a few elites that influence economic policy, is reminiscent of yesteryear.

Similarly, economic protectionism is also seen in the extraterritorial nature of proposed digital regulations. For instance, the NPD Committee states that "those who take Indian community or public Non-Personal Data outside India will bear full legal responsibility for complying with any such immediate or future data sharing or other regulatory requirements" (MeiTY 2020). This implies that digital businesses with a global footprint would need to comply with local regulations, that are not linked to protection of any fundamental right of citizens such as individual privacy. This conception is antithetical to any global efforts to create rules for digital highways. And in this sense, it also militates against India's commercial and strategic interests, in favour of common rules that can allow companies to scale in the global digital economy, without encountering unique regulatory constraints in every jurisdiction. Such barriers would be especially onerous for smaller digital businesses, because of high compliance costs. At the same time, such businesses would have the most to gain from a diversified consumer base and a chance to serve the global market.

The principle of data sovereignty exemplifies how modernisation and technological change do not necessarily lead to more enlightened policymaking. In 1991, India managed to unshackle itself from a so-cialist economic model characterised by centralised planning and an industrial policy approach to the development of a fledgling private sector. This model resulted in non-competitive industrial markets led by a few monopolists, heavy-handed economic regulation which en-trenched monopoly interests, and rampant discrimination against foreign businesses including nationalisation of their private property. Consequently, India had its back against the wall at the turn of the 1990s, with a rapidly deteriorating economy and a balance of pay-ments crisis, both symptoms of the lack of economic competitiveness.

Unfortunately, the digital economy has not led to a break from the past. It has instead encouraged some of the most regressive impulses within the policy establishment, rather than triggering an appetite for modern methods of governance that rely on greater accountability and transparency of all stakeholders in digital markets, including the State. Much of this can be traced back to the conception of data as oil, which suggests that all India has to do, is to regulate its extraction rather incentivise its refinement.

Data, Davos and false hope

India has turned to a form of digital nationalism, to unlock the po-tential it sees in the gold rush for data. Notions of data as oil or data sovereignty are a symptom of this paradigm. Paradoxically, the country has reversed a trend of three decades of economic liberal-isation, through the active pursuit of such misguided concepts in the digital sphere. The twist in the tale is also an unintended outcome of the impact of the sense of hope associated with India's embrace of globalisation, in the mid-2000s. Many expected India to become an economic superpower, because of the size of its young workforce. This reasoning is remarkably similar to the deterministic logic of inevitable dominance of the global digital economy because of heavy data con-sumption in the domestic economy.

In 2006, the erstwhile director-general of the Confederation of Indian Industry, a large association that represents industry interests, highlighted that India was often described as "closed and insular" before the nineties (Wharton 2006). Narasimha Rao was the first Indian Prime Minister to visit Davos in 1992, where he stated the mantra "change with continuity" (Mohan 2020). That is, India would liberalise further, but without sacrificing its public sector and while maintaining financial and regulatory prudence. A few years later, in 1997, H.D. Deve Gowda became the second Indian Prime Minister to visit the Summit. He recalled that "the general feeling about our country was that it was very corrupt" and that he "worked towards changing that image" (Swamy 2018). By the mid-2000s however, there was a discernible shift in perception because developing countries like India became engines of economic growth, through the combination of globalisation and a low base effect.

In 2009, the WEF hosted a debate asking "can India become a superpower", with the qualification that "experts predict that India will become a global superpower within fifty years – thanks to its economic growth and young population" (WEF 2009). Similarly, a 2010 New York Times article noted that "representatives from developing nations have been present in Davos for decades, but until recently they were often treated as aid recipients rather than partners in the global economy" (Bansal 2018). In 2018, Narendra Modi even championed globalisation at Davos. In his plenary speech, he said that protectionists wish to "change the natural flow of globalization". His remarks were seen as a "welcome contrast" to growing American isolationism under President Donald Trump (Bansal 2018).

Accounts of transformation of the developing world through globalisation were mirrored in the paradigm shifts associated with digitalisation, and therefore the notion that data has innate value. In 2012, a prominent market intelligence firm noted that the amount of data in the world coming from emerging markets will grow from 36 percent to 62 percent in 2020 (IDC, 2012). This growth potential led the head of the Business Software Alliance, a global association of technology companies, to proclaim that "millions of lives will be changed" through investments in data markets "in the developing world to address crucial issues" (Espinel 2017). She also noted that the notion that "deficiencies in infrastructure and resources prevent people in developing countries from using data and analytics to their advantage" is outmoded and that its time to "debunk the myth that people in these countries aren't ready to harness the power of data" (Espinel 2017).

India bought into the transformative potential of data hook, line and sinker, by embracing the notion that data is indeed the new oil. It did so without adequate efforts to address underlying challenges to industrialisation discussed earlier. In fact, the notion fostered bad policy design that steers the country towards industrial policies of yesteryear. As a result of the international narrative, India also saw itself as a future superpower and became complacent about addressing structural economic deficits such as the lack of infrastructure and institutional weaknesses. Consequently, it was among the worst impacted economies in the COVID-19 pandemic in 2020. In August 2020, there was consensus among the nation's top economists, including a former Prime Minister, that the country would face its first-ever economic contraction in nominal terms since Independence (Biswas 2020). This led a prominent public intellectual, to remark on India's seventy-third Independence Day, that on the "sixtieth anniversary of Indian Independence ... one could at least debate whether our country was a superpower in the making" and "now, 13 years later, such a debate would be farcical in the extreme" (Guha 2020).

Unless India course corrects and invests in charting out blueprint for a globally competitive digital ecosystem, with compatible guiding principles, its self-conception as a modern digital superpower may be similarly short-lived. One final indicator may help substantiate this view. In 2019, India had among the highest per capita mobile data consumption figures in the world; however, less than 10 percent of website requests were served locally, compared to 74 percent in US-Canada, and 42 percent in East Asia including China (UNCTAD 2019). That is, most website requests are routed to servers abroad. India must enlarge its global footprint, by not repeating past mistakes. This includes the critical mistake of interpreting an opportunity like digitisation to be somehow self-fulfilling.

India's IT industry is built on cheap labour, and the data is oil construct is analogous because it assumes an opportunity to build a digital industry based on cheap data. From the late eighties onwards, India's large, young engineering workforce was deployed to work in back-office services. These jobs were primarily outsourced by the West, because of the comparative cost advantages. But like cheap labour, cheap data is an ephemeral advantage. Smaller countries like the Philippines and Vietnam have started to chip away at India's share of the outsourced IT market (Patel 2018). Similarly, despite high mobile data consumption, only a handful of digital companies have monetised this demand. Behind this paradox of plenty is the fact that India's purchasing power has not kept up with its superpower ambition, and the latter has contributed to a vicious cycle of poor policy planning and inertia. Moreover, principles like data sovereignty are antithetical to a competition imperative, because the protectionism associated with them creates barriers to real competition and disincentives to innovation. Why should Indian digital businesses spend money on innovations that can serve local consumers better, if the regulatory framework encourages incumbency? Why should these businesses feel the need to venture outside India and compete with global businesses, if the local economy can remain a protected turf? Why risk uncertain returns on research and development, if cheap data combined with protectionist regulation assures fixed returns?

Moreover, the bipolar dominance (American and Chinese) of the global digital and technology economy will remain unchallenged, without a blueprint for the design and development of Indian products and platforms. In 2020, strategist C. Raja Mohan wrote that although "Delhi ought to be at the leading edge of the current debate on the future of capitalism" the country "seems too preoccupied sorting out the persistent legacies of feudalism" (Mohan 2020). Some of these preoccupations are visible in the digital sphere – where a legacy control mindset persists. These manifest in various forms – licencing, price controls and expropriation of private property. The conception of data as oil only reinforces legacy rulemaking. And while the principle of data sovereignty opposes digital imperialism, it employs colonial-era policy tools to exercise state control. The irony is discomforting and highlights the opposite paths of modernity as represented by digital markets and economic policy.

Notes

1 Smith explained these factors through practical examples. For instance, he saw the price of wheat as a sum of the rent of a landlord, the wages of labourers and the profit of the farmer who provides money and equipment to operate the business.
2 See: Lucas 1988, Makiw et al., 1992.
3 2019 list of the top 1000 of the World's Banks, according to The Banker Database.

Bibliography

Aiyar, Swaminathan S. Anklesaria. "India's New Protectionism Threatens Gains from Economic Reform." Policy Analysis Number 851, The CATO Institute, 2018. https://www.cato.org/policy-analysis/indias-new-protectionism-threatens-gains-economic-reform

"Average Cost of Factory Labour At Less Than $2 Per Hour Gives India Big Advantage of Wage Arbitrage: Bain And Co. Worldwide Managing Partner, Manny Maceda." *The Economic Times*, April 03, 2018. https://economictimes.indiatimes.com/opinion/interviews/average-cost-of-factory-labour-at-less-than-2-per-hour-gives-india-big-advantage-of-wage-arbitrage-bain-and-co-worldwide-managing-partner-manny-maceda/articleshow/63554253.cms?from=mdr. (accessed on April 18, 2021).

Bansal, Paritosh. "Indian PM Modi Defends Globalization at Davos Summit." *Reuters*, January 23, 2018. https://www.reuters.com/article/us-davos-meeting-modi/indian-pm-modi-defends-globalization-at-davos-summit-idUSKBN1FC1AL. (accessed on April 25, 2021).

Biswas, Soutik. "Manmohan Singh's 'Three Steps' To Stem India's Economic Crisis." *BBC*, August 10, 2020. https://www.bbc.com/news/world-asia-india-53675858. (accessed on April 25, 2021).

"Can India Become a Superpower?." *Davos Debates*, World Economic Forum, October 21, 2009. https://www.weforum.org/agenda/2009/10/can-india-become-a-superpower/. (accessed on April 25, 2021).

Chaudhary, P.P. "Benefits of Digital Economy." *Lok Sabha Unstarred Question* No. 2944, August 02, 2017. http://164.100.24.220/loksabhaquestions/annex/12/AU2944.pdf

"Cinema Screens in India Woefully Low, Hitting Global Ranking." *Times of India*, June 18, 2017. https://timesofindia.indiatimes.com/business/india-business/cinema-screens-in-india-woefully-low-hitting-global-ranking/articleshow/59201818.cms. (accessed on April 25, 2021).

"Corruption in the Extractive Valuechain: Typology of Risks, Mitigation Measures and Incentives." OECD Development Centre, 2016. https://www.oecd.org/dev/Corruption-in-the-extractive-value-chain.pdf

"Data Is the New Oil, the New Gold": PM at 'Howdy, Modi! "In Houston." NDTV, September 23, 2019. https://www.ndtv.com/india-news/data-is-the-new-oil-the-new-gold-says-pm-modi-in-houston-2105338. (accessed on April 18, 2021).

"Delhi in Davos: How India Built its Brand at the World Economic Forum." Knowledge @ Wharton, University of Pennsylvania, February 22, 2006. https://knowledge.wharton.upenn.edu/article/delhi-in-davos-how-india-built-its-brand-at-the-world-economic-forum/

"Digital Economy Report." United Nations Conference on Trade and Development, 2019. https://unctad.org/system/files/official-document/der2019_en.pdf

"Ericsson Mobility Report." November 2019. https://www.ericsson.com/en/press-releases/2/2019/11/ericsson-mobility-report-5g-subscriptions-to-top-2.6-billion-by-end-of-2025

"Ericsson Mobility Report." November2021.. https://www.ericsson.com/en/reports-and-papers/mobility-report/reports/november-2021.

Espinel, Victoria. "The Power of Data in the Developing World: Making the Software Revolution Global." *Economic Times*, May 02, 2017. https://cio.economictimes.indiatimes.com/tech-talk/the-power-of-data-in-the-developing-world-making-the-software-revolution-global/2326. (accessed on April 25, 2021).

"Four Ways the NSA Revelations Are Changing Businesses." *The Guardian*, June 09, 2014. https://www.theguardian.com/media-network/media-network-blog/2014/jun/09/edward-snowden-nsa-changing-business. (accessed on April 25, 2021).

Guha, Ramachandra. "Thirteen Years on, Preening Over the 'India Story' Has Been Replaced by a Sense of Gloom." *The Scroll*, August 16, 2020. https://scroll.in/article/970454/ram-guha-thirteen-years-on-india-story-preening-has-been-replaced-by-a-sense-of-gloom. (accessed on April 25, 2021).

Hermes, Sebastian et al. "Breeding Grounds of Digital Platforms: Exploring the Sources of American Platform Domination, China's Platform Self-Sufficiency, and Europe's Platform Gap." *Association for Information Systems*, Research Paper 132, 2020. https://aisel.aisnet.org/ecis2020_rp/132/

"Indian IT Should Move to Intellectual-Arbitrage Model: Cognizant's R Chandrasekaran." *Economic Times*, January 25, 2019. https://economictimes.indiatimes.com/corporate/indian-it-should-move-to-intellectual-arbitrage-model-cognizants-r-chandrasekaran/articleshow/67680254.cms. (accessed on April 25, 2021).

"India's China Envy.", NPR Special Series, May 20, 2010. https://www.npr.org/templates/story/story.php?storyId=127014493

"India Can Play Pivotal Role in Global Fourth Industrial Revolution." *Economic Times*, April 11, 2018. https://cio.economictimes.indiatimes.com/news/strategy-and-management/india-can-play-pivotal-role-in-global-fourth-industrial-revolution/63707408. (accessed on April 25, 2021).

Johari, Sneha. "Indian Data Should Be Owned by Indians, Not Corporations – Mukesh Ambani, RIL." *Medianam*, January 21, 2019. https://www.medianama.com/2019/01/223-india-data-localisation-mukesh-ambani/. (accessed on April 25, 2021).

Jones, Steve. "Why Big Data Is the Fourth Factor of Production." *Financial Times*, December 27, 2012. https://www.ft.com/content/5086d700-504a-11e2-9b66-00144feab49a. (accessed on April 25, 2021).

Kumar, Shailender. "Indian IT 4.0: Upping the Ante on Innovation." *Economic Times*, November 30, 2019. https://economictimes.indiatimes.com/tech/ites/indian-it-4-0-upping-the-ante-on-innovation/articleshow/72303400.cms?from=mdr. (accessed on April 25, 2021).

Lucas, Robert E. "On the Mechanics of Economic Development." University of Chicago, February 1988. https://www.parisschoolofeconomics.eu/docs/darcillon-thibault/lucasmechanicseconomicgrowth.pdf.

Makiw, Gregory N., David Romer , and David N. Weil. "A Contribution to the Empirics of Economic Growth." *The Quarterly Journal of Economics* 107, no. 2 (May 1992). https://academic.oup.com/qje/article-abstract/107/2/407/1838296

Ministry of Electronics and Information Technology, Government of India (MeiTY), Report by the Committee of Experts on Non-Personal Data Governance Framework, 2020. https://static.mygov.in/rest/s3fs-public/mygov_159453381955063671.pdf

Ministry of Commerce and Industry, Government of India, Draft National e-Commerce Policy, February 23, 2019. https://dipp.gov.in/sites/default/files/DraftNational_e-commerce_Policy_23February2019.pdf

Mishra, Asit Ranjan."The Idea of a Universal Basic Income in Perpetuity Is Unjust." *The Mint*, April 01, 2019. https://www.livemint.com/news/india/the-idea-of-a-universal-basic-income-in-perpetuity-is-unjust-rathin-roy-1554055673305.html. (accessed on April 25, 2021).

Mittal, Shivangi, and Ramdas, Varun. "Indian TV Broadcasting at a Crossroads." Koan Advisory Group, August 2020. https://www.koanadvisory.com/wp-content/uploads/2020/08/Indian-TV-Broadcasting-at-a-Crossroads-2.pdf

"Modi Announces Package for Land and Labour, Talks About Make in India 2.0 Amid Coronavirus Crisis." *Business Today*, May 13, 2020. https://www.businesstoday.in/current/economy-politics/modi-announces-package-for-land-and-labour-talks-about-make-in-india-20-amid-coronavirus-crisis/story/403652.html. (accessed on April 25, 2021).

Mohan, C. Raja. "Delhi-Davos Disconnect—India Must Find Ways To Take Advantage Of New Opportunities." *Indian Express*, January 21, 2020. https://indianexpress.com/article/opinion/columns/davos-world-economic-forum-trump-india-6226839/. (accessed on April 25, 2021).

Mohan, C. Raja "India Must Reflect Frankly upon China's Extraordinary Transformation." *The Indian Express*, October 1, 2019. https://indianexpress.com/article/opinion/columns/china-peoples-republic-70th-anniversary-6042685/. (accessed on April 18, 2021).

Mohan, Archis. "Congress Finally Remembers Narasimha Rao On 29th Anniversary Of 1991 Budget." *Business Standard*, July 24, 2020. https://www.business-standard.com/article/politics/congress-finally-remembers-narasimha-rao-on-29th-anniversary-of-1991-budget-120072401146_1.html. (accessed on April 25, 2021).

Pahwa, Nikhil. "Nationalisation of Data Will Destroy Value for Businesses, Investors." *Times of India*, August 02, 2020. https://timesofindia.indiatimes.com/blogs/toi-edit-page/nationalisation-of-data-will-destroy-value-for-businesses-investors/. (accessed on April 25, 2021).

Palmer, Michael "Data is the New Oil." Blog Entry, ANA Marketing, November 3, 2006. https://ana.blogs.com/maestros/2006/11/data_is_the_new.html. (accessed on April 25, 2021).

Patel, Deepak. "IT Industry Facing Tough Competition from Vietnam, Philippines." *Indian Express*, April 12, 2018. https://indianexpress.com/article/business/business-others/it-industry-facing-tough-competition-from-vietnam-philippines-5133678/. (accessed on April 25, 2021).

Peermohamed, Alnoor. "Covid Impact: 70% Of Startups Have Cash Reserves to Last Less Than 3 Months." *Economic Times*, May 19, 2020. https://economictimes.indiatimes.com/small-biz/startups/newsbuzz/indias-startups-story-hanging-by-a-thread/articleshow/75817730.cms. (accessed on April 25, 2021).

Perez Sebastian, Fidel, Ohad Raveh and Frederick van der Ploeg. "Oil Discoveries and Protectionism: Role of News Effects." *Journal of Environmental Economics and Management*, 2021. 10.2139/ssrn.3379050

Rodrik, Dani. "Premature Deindustrialisation." *IAS School of Social Science*, Paper Number 107, January 2015. https://drodrik.scholar.harvard.edu/files/dani-rodrik/files/premature-deindustrialization.pdf

Sengupta, Jayshree. "Is India Prepared For Fourth Industrial Revolution?." India Matters, Observer Research Foundation, May 10, 2017. https://www.orfonline.org/expert-speak/is-india-prepared-for-fourth-industrial-revolution/

Sherman, Justin. "India's Data Protection Bill in Geopolitical Context." Blog Post, New America Foundation, July 10, 2019. https://www.newamerica.org/cybersecurity-initiative/c2b/c2b-log/indias-data-protection-bill-geopolitical-context/. (accessed on April 25, 2021).

Singh, Gurbachan. "Land In India: Market Price Vs. Fundamental Value (blog)." Ideas for India, February 29, 2016. https://www.ideasforindia.in/topics/governance/land-in-india-market-price-vs-fundamental-value.html

Sinn, Hans-Werner. "Social Insurance, Incentives and Risk Taking." Working Paper 5335, National Bureau of Economic Research", 1995. https://econpapers.repec.org/paper/nbrnberwo/5335.htm

Swamy, Rohini. "Last PM To Go to Davos Says Modi Shouldn't Take All Credit for India's Progress." The Print, January 24, 2018. https://theprint.in/report/pm-davos-modi-credit-india-progress/31410/. (accessed on April 25, 2021).

"The Digital Universe in 2020: Big Data, Bigger Digital Shadows, and Biggest Growth in the Far East." IDC, December 2012. https://assets.ey.com/content/dam/ey-sites/ey-com/en_gl/topics/digital/idc-the-digital-universe-in-2020.pdf

Yakunina, R.P., and G. Bychkov. "Correlation Analysis of the Components of the Human Development Index Across Countries." *Procedia Economics and Finance*, 2015. https://www.sciencedirect.com/science/article/pii/S2212567115006929

5 Digital in the Time of COVID

The COVID pandemic brings to the fore the different puzzling paradigms we have explored over the course of the book. It brings home the fact that while technology can accomplish many erstwhile impossible feats, it cannot bridge critical gaps in policy design or stand in for the State in core areas of governance such as accountability of public service delivery, arguably where brick-and-mortar institutions are most needed. Rather technology, particularly digital, often serves to exacerbate and complicate governance challenges. Simultaneously, the greater the reliance on digitalisation, the greater the unwillingness to engage with messier elements of governance that technology itself cannot account for. In the midst of a crisis, such as the pandemic that befell India and the world in 2020, the consequences of such characteristic correlations are somewhat dire.

COVID was first discovered in Wuhan, a city in the Hubei province of China, in late 2019. Reports surfaced of a new kind a pneumonia whose origin was unknown. Twelve days later, the disease claimed its first reported victim in China, a 61-year-old man (Gale 2020). Shortly thereafter cases were confirmed in Europe and the United States. The Indian Ministry of Health and Family Welfare (MoHFW) issued an advisory stating that it would screen passengers travelling back to India from China at the airport. India's first reported case of coronavirus came on 30 January 2020; a man in the Thrissur district of the southern state of Kerala who was previously studying in Wuhan (Kumar 2020). Between 9th and 12th January, the World Health Organisation (WHO), a multilateral institution that focusses on public health, issued a series of documents for countries on strategies for the management of the COVID outbreak. Key among these was the "National Capacities Review Toolkit", that provided a framework for assessment of national preparedness for when the disease struck (WHO 2020). The document highlighted the various facets of outbreak

DOI: 10.4324/9780429324901-5

response strategy that must be deployed in tandem for the successful management of the disease both before and after it permeates a country's borders. Surveillance formed a key component of this management plan. Several nations, including India, deployed technological tools such as mobile applications to aid health surveillance efforts.

In this chapter, we first explore India's experience with applications developed for contact tracing to manage and control the sudden outbreak of COVID. In particular, in doing so, we attempt to highlight the correlations between lack of focus on critical elements of technological interventions such as trust and good design. This is a recurrent theme in this book, and is one that helps to illustrate why the dazzle of digital can sometimes blind side decisionmakers and generate paradoxical outcomes. We subsequently detail India's experience of navigating the social welfare nightmare that was COVID. The country attempted to reach cash to the poor through the use of digital financial services built on top of the much-celebrated unique identity programme, Aadhaar. The perverse outcomes of this exercise in the early months of March–May 2020 illustrate that while digital is made out to be a panacea for hard challenges of effective developmental interventions that can benefit the poorest of the poor, it can sometimes exacerbate the divide between the haves and the have nots. In sum, this chapter builds on the arguments of previous chapters, only in the context of a healthcare emergency, where one can argue technology was most needed, and yet, failed.

Technology as the sole means for remote surveillance

Surveillance is an important element of managing viral outbreaks. It enables both domestic and international authorities to detect and monitor the spread of disease and chalk out effective solutions for case management (WHO 2006). The early warnings afforded by surveillance are crucial for health security (WHO 2006). Further, it can help medical experts identify the long-term effects of a disease, such as in the case of the Zika virus fever (Gallagher 2018). The International Health Regulations (2005), to which India is a signatory, also mark the importance of a well-equipped surveillance apparatus in the battle against communicable diseases. Article 5 of the Regulations requires States to develop, strengthen and maintain the capacity to detect, assess, notify and report events (IHR 2005). Surveillance has been useful in aiding the management and mitigating the spread of disease in India as well. The WHO's polio surveillance network, for instance, played a key role in eradicating the disease from the country. In the case of COVID, the WHO reiterated that "robust surveillance" was critical to "control the spread" of the disease and guide the "implementation of control measures", "detect and contain outbreaks among vulnerable populations", study the effect of the outbreak on healthcare systems and societies, and observe its co-existence with other similar diseases in a population (WHO 2020).

India set up its disease surveillance system, the Integrated Disease Surveillance Programme (IDSP), in 2004, with support from the World Bank. The IDSP was established to monitor disease trends and facilitate quick responses to outbreaks. It had surveillance units at the central, state and district levels in India. Such decentralisation enabled it to monitor and report on disease outbreaks throughout the country. These reports were issued weekly and were made available on the IDSP online portal. For COVID, the surveillance mandate, along with the overall task for managing the outbreak, was shifted to the Indian Council for Medical Research (ICMR). Ostensibly, the break from the status quo was for several reasons. One, the Government decided to give primacy to ICMR because the expeditious approval of diagnostic kits was required as they were dealing with a new disease. Two, the IDSP was comparatively underfunded. In 2017–2018, IDSP's spending amounted to USD 4.1 million whereas the ICMR spent USD 185 million. Three, the ICMR enjoyed relatively greater autonomy than the National Centre for Disease Control (NCDC), the agency within which the IDSP is housed. The head of the ICMR, the Director General, is also the Secretary of the Department of Health Research under the MOHFW. Fourth, and somewhat relatedly, the ICMR is also more prominent than the NCDC, as it engages directly with the Health Minister. Fifth, the IDSP was also understaffed. In March 2020, the MoHFW put out an expression of interest for hiring six epidemiologists for the Central unit of the IDSP. In April 2020, it was reported that 216 out of 736 districts did not have epidemiologists. The deficit in trained personnel was highlighted in 2018 as well by the WHO in its Risk Assessment on the NIPAH virus outbreak in India. The agency suggested that the dearth in human capital compelled a significant amount of improvisation in the early response of the government to the epidemic. Therefore, there were large gaps in state capacity that existed pre-COVID, and the shift to ICMR was at best a palliative measure. In fact, despite the change in management to ICMR, advisories issued around data collection and reporting at airports indicated that much of the on-ground surveillance was still carried out by the short-staffed IDSP.

A dearth of personnel coupled with the requirement for social distancing prompted a reliance on technology for remote surveillance. This was a logical step for India in the effort against the virus. Digital technology, has in the past, had a positive impact on the management of an epidemic (Bassi et al. 2020). Mobile health applications proved useful in battling the Zika and Ebola epidemics by aiding efforts towards contact tracing and testing (Bassi et al. 2020). Several countries deployed mobile health applications as part of their COVID-19 response. These included Russia, China, South Korea and Singapore.

Aside from global best-practice, technological intervention is a good signalling tool – an easy way to indicate a forward-thinking approach to citizens – of a State that is ready to deploy innovative solutions to tackle novel challenges. The benefits of such signalling, however, only go as far as its utility to handling the issue at hand. As a consequence, when technology solutions are linked to flagship government programmes, there is always a risk that overemphasis on signalling may overshadow important considerations, such as the building citizens' trust or designing interventions well – both key determinants of success.

Aarogya Setu, the State's flagship COVID-surveillance application, was reportedly downloaded 127 million times (Verma 2020). Even with the supposition that each download corresponds to a separate individual, that is only one-tenth of the population. According to reports, in order to be effective, contact tracing apps must be downloaded by roughly 50 to 70 percent of the population (Akinbi et al. 2021). A seroprevalence survey carried out by ICMR in May–June 2020 indicated that for every reported infection between 80 and 130 were missed (Garari 2020). These figures are perhaps the starkest indictment of the State's effort to manage COVID, despite the use of technology.

Designed for mistrust

Trust is important for the introduction of a technological solve in a heterogenous society like India (Colesca 2009). For a technological tool to be effective, people need to use it. For people to use new tools, they need conviction about its safety and convenience – which cannot be built remotely or through communications campaigns alone.

There are multiple ways in which such a trust paradigm plays out. First, there is the trust between citizens and the State. Despite India's sizable digital footprint, most individuals are still not adept at dealing with digital applications. This demographic must be convinced about why a digital application is important for their well-being. Then, there are those who are technologically savvy, but are concerned about matters such as surveillance and privacy. Their confidence may be won through robust terms of use and privacy policies that describe how their data will be used by those who control the application.

Meaningful efforts to garner citizens' trust were absent in the scheme of COVID-surveillance applications in India. For instance, there was a consistent dearth of transparency surrounding the launch of the primary COVID-surveillance application, "Aarogya Setu". It was originally positioned as a contact tracing app that used Bluetooth to indicate when one user was close to another. However, subsequent reports indicated that an official committee was set up to monitor the data gathered by the app for "determining whether India should relax lockdown conditions" and possibly "initiate geo-fencing mechanisms" (Deb 2020). It was also reported that the State may use the app to punish those that violated quarantine orders. Privacy advocates also noted that there was no way of knowing whether the data gathered by the app was adequately anonymized. These reports, coupled with questions on rights and civil liberties, prompted a mistrust of its intentions for the data generated by the application.

Further misgivings were stoked when state agencies mandated the Aarogya app for certain sections of the population. The Ministry of Home Affairs, for instance, issued guidelines that mandated all businesses that were operating to ensure their employees were using the application. Police in a locality in North India even suggested that individuals could be charged for not having the app on their phones. People in Delhi with smartphones were not allowed to enter police buildings if they did not have the app installed (Singh 2020). To be clear, India was not the only country trying to mandate health surveillance through such applications. Other mandatory programmes were carried out in countries like China, South Korea, and Israel in the early aftermath of the global outbreak, while several others like Australia and United Kingdom developed official applications for voluntary use.

Security issues were another concern that led to erosion of trust at the level of citizens. An ethical hacker with the pseudonym Elliot Alderson highlighted on social media that the Aarogya app had a glitch that allowed access to any internal file on it. Essentially, the hacker could tell if a person was sick at the most important and sensitive of government offices, such as the Prime Minister's Office. Though the State denied the presence of any such security gap, the hacker alleged that it quietly fixed the problem after he pointed it out.

A second reason for the failure of technological interventions is the lack of trust between political actors. One hopes that at a time of national calamity, political parties will leave their differences aside and come together to help fight a crisis. The polarised politics of our times, however, preclude such an outcome, and India is no exception. Any disaster, particularly at the national level, is seen more as an opportunity for parties to rack up points against one another than an occasion for bipartisanship.

Under the Indian Constitution, health is a state subject which means that each state may govern its health apparatus independently. States that were not governed by the BJP, the ruling party at the Centre, exercised their Constitutional autonomy to manage the pandemic. This divergence included technological deployment. Thus, all of them, namely Kerala, Maharashtra, Delhi, Tamil Nadu, Rajasthan and West Bengal launched their own applications or integrated a COVID function into an existing local e-governance app for data gathering and disease surveillance (Bassi et al. 2020). The problem with a response that is fragmented along political lines is that it frustrates remote surveillance efforts to a great extent. With so many digital applications, many of which carry out similar functions, datasets were splintered and fragmented. Reports reveal how the limited centre-state coordination and lack of trust meant that data was not being shared, perhaps for fear of leakage of politically sensitive COVID statistics. For instance, the Union Minister of State for Health and Family Welfare, Bharati Pawar, noted that states did not share data on Covid-related deaths caused by oxygen shortage (Hindustan Times 2022).

Several state authorities in India had an interest in reporting low infection rates. This is because it became easy to politicise the rise in infections by directly attributing it to the mismanagement of the disease at the state level. Conversely, the rate of infection could be presented as directly proportional to the adequacy of steps taken to fight the spread of the disease. This impulse to control statistics was evident from the time COVID first broke. The ICMR released a set of restrictive testing guidelines in March 2020. Only government facilities were allowed to carry out COVID tests. Patients with a travel history from Wuhan within the prior two weeks and symptoms of acute respiratory distress were the only ones that could be tested. The restrictive nature of these guidelines was attributed to the paucity of tests and the relatively low reported incidence of the disease in the country. Another concern was the lack of adequate health facilities that were certain to be overwhelmed if there was a rush for testing or admission. This made sense, especially since the disease was mild in a majority of cases and capacity needed to be reserved for those with more severe symptoms. However, the bar on allowing private facilities to test indicates that politics was at play as well. Officials may not have wanted the pandemic to highlight yet another instance where the most vulnerable, such as poor migrants, were denied access to essential facilities that were available to the wealthy. Further, restricted testing meant fewer infections reported, which, at least superficially, indicated that it was doing a good job of handling the pandemic.

Finally, there was the lack of trust between the State and private healthcare service providers, despite the shortfall in public service delivery of health. Health has never been a top policy priority in India. Over the last decade, the country invested around two percent of its GDP in health, more than seven percentage points below the global average in 2021 (OECD 2021). India's health infrastructure suffers from a severe lack of funds and is in a state of disrepair. Private healthcare is relatively better off. However, there are important caveats. One, quality private healthcare is concentrated in urban centres. Two, the private sector is not homogeneous even in urban areas. There are different types of service providers that vary in both quality and capacity. Three, and most importantly, private health facilities are not beholden to the same set of priorities as those in the public sector, namely social policy or welfare. It may be argued that the Hippocratic Oath taken by doctors should prompt the profession to shun the pursuit of profit in the quest of the more noble goal of saving lives. But the pursuit of profits can sometimes contradict the spirit of the Oath, particularly in emergency situations when windfall gains can be made. Despite these considerations, governments at the Centre and states, placed great onus on the private health sector to play its part to support the fight against COVID. Several hospitals across the country, 600 at last count, were designated COVID hospitals. The reassignment of these facilities to COVID-care, as one doctor noted is a big task (*Hindustan Times* 2020). Converting a hospital into a COVID facility requires "structural separation" – breaking down walls, building new ones to physically segregate COVID patients from non-COVID ones (Hindustan Times 2020). Air-conditioning units must be compatible, with at least 12 air circulations an hour for an area to be COVID compliant (Hindustan Times 2020). There cannot be a simple requisitioning of capacity or sequestering of physical space.

Despite the clear moral imperative to cooperate with these facilities, however, certain state governments such as that of the nation's capital Delhi, issued a series of arbitrary diktats to requisition private healthcare facilities for the fight against COVID. Illustratively, the Delhi Government passed an order in May directing 117 private hospitals and nursing facilities to reserve 20 percent of their beds for COVID

(Dutt 2020). Thereafter, on 2 June 2020, the Delhi Government mandated that only symptomatic patients could be tested. This was done despite widespread reports of asymptomatic infections.

The dearth of trust between public and private institutions possibly reached its most absurd apogee when the Delhi Government filed criminal charges against Sir Gangaram Hospital, because it neglected to supply data to the former's RT-PCR testing application. The RT-PCR application was launched by the Delhi Government for doctors to enter sample collection details for COVID. Doctor Ambarish Satwik, a vascular surgeon at the hospital in question, stated in an interview that the hospital was sharing information with the Integrated Disease Surveillance Portal, a digital repository setup by the IDSP for online disease reporting. He further remarked that the authority that filed the charge sheet against the hospital acknowledged this fact. Yet, the Delhi Government saw the need to go ahead and file a criminal case against it, possibly because the former saw the reporting failure as wilful disobedience against its authority.

The Gangaram Hospital experience highlights another major failing in the deployment of technology by the State during the pandemic, namely poor design and implementation. One design problem was the deluge of apps. A study by Bassi et al. (2020) found that there were 346 potential COVID applications in the country. Beyond the dearth of trust between competing political interests that prompted a multiplicity of applications, there was also a lack of coordination between different wings of the same governments. For instance, at the central level, there were multiple apps and portals doing the same thing. Aside from the RT-PCR app, there was an IDSP portal for data collection, as well as a portal residing with ICMR for medical staff to log information about testing samples. Similarly, the state of Karnataka developed a contact tracing app called Corona Watch, despite being ruled by the same party that was sitting in power at the Centre.

A second design problem was application accessibility. The contact tracing app Aarogya Setu was only offered in 12 languages, despite the existence of at least 22 distinct spoken dialects, recognised by the Indian Constitution. Moreover, several of the data entry apps, including the Delhi government's RT-PCR app, required medical staff to register via a text led two-factor authentication system. This added a layer of friction in the data entry process as many individuals failed to receive their one-time passwords due to poor connectivity. For instance, in Kerala, the GoK Direct Kerala Quarantine Watch, which required users to enter their zip code to login, failed to recognise many of the districts including prominent ones such as Thiruvananthapuram.

The design flaws in the applications launched by central and state governments speak to the larger problem in the way the State wields technology. In each instance, technology was deployed as a top-down solution. The State relied on technology as a solutionist overlay that will neatly insert itself into broken systems to plug the cracks within them. The IDSP was short-staffed. Thus, an application was launched to gather data directly from citizens. Rather than build public confidence in the utility and necessity of the app and give assurances around the sanctity of its data sharing practices, the State resorted to coercive action to prompt individuals to download it. Similarly, multiple applications and portals were created to ensure that limited resources such as beds and other medical facilities were optimised and the State had real-time insights on the situation on the ground. These interventions were, of course, not consonant with the reality on the ground where hospitals were dealing with staff shortfalls and a deluge of patients. In such circumstances it is likely that doctors would not have the bandwidth to engage in multiple forms of repetitive data entry.

The multiplicity of applications and the lack of coordination and efforts to build confidence between the holders of different datasets meant that information was incomplete and those responsible for data entry, doctors and individual citizens, faced app fatigue. Consequently, at the peak of both the first and the second wave, information on the availability of hospital beds was incomplete. Several citizens undertook their own data collection, calling hospitals and creating public spreadsheets on the status of availability of different medical services. We, the authors, were also engaged in such activities, trying to source oxygen cylinders and concentrators, and hospital beds for those who reached out to us. Citizens relied primarily on technology to disseminate information about the availability of resources. However, there was personal engagement on the ground, largely in the form of phone calls to different facilities, to verify the information circulated. The initiative taken by citizens, especially in the face of second wave, attests to the power of technology as a force for good when combined with the best of humanity as well as a searing indictment of how it is used by the State to abdicate its responsibility.

India's experience with technology in the fight against COVID underscores the rise of a systemic challenge in society, one that took root well before the arrival of the pandemic, namely the refusal of the State to view itself as separate from the technological apparatus put out by it. The State views any technological intervention as an extension of itself. At a superficial level, the political logic of such a notion is seemingly sound. The virality with which technology can spread makes it more effective in garnering the goodwill of the polity than any welfare scheme in the physical realm. This is logical for governance as well, since digital allows for the transcendence of the physical, such as a lack of infrastructure or resources. However, placing so much faith in technology or an app, when only half of the population has smart-phones, is necessarily flawed. Additionally, given the quality of connectivity, not to mention the low rates of literacy in most rural areas across the country, there was limited hope of success through such a measure even at the outset. Yet the insistence on choosing such an avenue, and then failing to focus on such key aspects as trust and design, reveals that technology will continue to serve as a convenient scapegoat for governance.

Battling covid with biometrics and data

The COVID pandemic also exposed the paradoxes of India's reliance on digital infrastructure to provide social security. The pandemic ravaged household incomes across the world, and engendered a need for large fiscal stimuli. The need for fiscal stimulus was particularly acute in developing countries where social security nets are thin, and most employment is a combination of temporary, contractual and informal. A high level of income and job fragility was laid bare in the initial few months – India lost around 121 million jobs in April (Vyas 2020). While a majority of these jobs recovered towards the end of the year, the dent in the income profiles of mot households is likely to drive a structural shift in per capita prosperity.

An early indication of the structural impact of COVID came in mid-2020 was when eminent economists like Carmen Reihart, a professor at the Harvard Kennedy School and a former deputy director at the IMD, and Kenneth Rogoff, also a Harvard professor and a former IMF Chief Economist began to speak out. They indicated that it may take up to five years for global per capita GDP to recover to 2019 levels, with a natural delay in the ability of developing countries to bounce back. Such warning bells weren't taken lightly. The chances of a persistent threat to economic growth prompted governments across the world to loosen their purse strings and kickstart some of the largest fiscal transfer programmes in history. A 190 countries and territories planned, introduced or adapted social protection measures in response to the pandemic by May 2020 (Gentilini et al. 2020). India was no exception.

India relied on its new digital infrastructural backbone to transfer cash directly to the hands of individuals and households during COVID. This infrastructure was built to leverage the unique identity programme called "Aadhaar", that became an integral part of the country's devel-opment assistance toolkit for domestic social welfare programmes. The unique ID was linked to millions of bank accounts and utilised as a tool for "Direct Benefits Transfer", or the transfer of cash to beneficiaries under various social schemes. Aadhaar and its sister programmes re-ceived equal praise as a transformative tool, from successive govern-ments from 2013 till today. In 2013, former Prime Minister Manmohan Singh summed up the potential of this Aadhaar-based social welfare ecosystem stating that "it will lead to better targeting of subsidies and reduce delays in the delivery of benefits such as scholarships and pen-sions to the intended beneficiaries. It will also help in curbing wastages and leakages, and result in greater financial inclusion" (Singh 2013). We explore this dual contention – of greater efficiency and financial inclu-sion that is repeated throughout the last decade or so in government speeches, in the context of the response to COVID. Both have figured in speeches of the current Prime Minister too, who has been unequivocal about the potential to achieve Antyodaya through Aadhaar – that is, reaching development programmes to the poorest of the poor with ease and efficiency (Prime Minister's Office, 2021). Indeed, the promise of a well-designed unique identification programme, linked to financial and banking services infrastructure and social security programmes is pre-cisely this. However, the devil lies in the details of such design, and those must stand the test of academic scrutiny.

Before we review academic literature on the flaws in the design of the financial inclusion scheme based on Aadhaar and recount how these came to the fore during COVID, it is useful to detail the aforementioned dual benefits of efficiency and inclusion. The claim that Aadhaar or any unique identity tool can increase the efficiency of social welfare pro-grammes, generally stems from the fact that these can eliminate inter-mediaries or middlemen. This is not dissimilar to the discussion in this book around the purported merits of digitalisation in agricultural markets. That is, intermediaries such as the dealers who provide rations under the Public Distribution Scheme for food grains, or postmen who deliver money orders to pensioners, are no longer required. This disin-termediation can potentially result in lesser corruption of the sort that was exemplified in former Prime Minister Rajiv Gandhi's now famous assertion that only 15 paise of every rupee meant for social welfare schemes ultimately reaches the intended beneficiaries. In fact, the Supreme Court of India even cross-referenced this rather candid ad-mission of the failures of the Indian State in achieving Antyodaya, in a judgement linked to a privacy challenge mounted against Aadhaar in 2018 (Rajagopal 2018).

The claims regarding the Aadhaar ecosystem's efficiency are also linked to a reduction in fraud. For instance, when Aadhaar numbers are linked to the Public Distribution System's database, a theoretical 100% coverage is achieved of beneficiaries with their entitlements. Even in the early stages of Aadhaar's social welfare ecosystem rollout, predating the pandemic, the Government "had claimed better targeting of beneficiaries and concomitant savings". The second major benefit of using Aadhaar as a vehicle for social welfare, is closely linked to the ability to better target beneficiaries – it is the hallowed objective of greater financial and developmental inclusion. Close to half of Indian citizens did not have a bank account until 2014, when Aadhaar-based seeding of bank accounts begun. Prime Minister Modi launched his flagship "Jan Dhan Yojana" in the same year, since when over 400 million new accounts have been created under this scheme (*Economic Times* 2020). This is a remarkable feat of preliminary financial inclusion, which has won the country global accolades and a Guinness Record!

So, the question is, what was the effectiveness of the Aadhaar-based fiscal transfer programme in the immediate aftermath of COVID, given that its efficiency and inclusion benefits were much touted? First, the global perspective. Current account transfers accounted for around a fourth of monthly GDP per capita across the 190 countries that implemented social security programmes in the immediate aftermath of the COVID outbreak (Gentilini et al. 2020). On average, such transfers more than doubled compared to average pre-COVID transfer levels. A similar surge was seen in India in the early months of March, April and May 2020. Around 330 million people were provided around INR 1000 each through financial assistance packages as on 22 April 2020 (Government of India 2020). However, researchers from Yale and the University of Southern California found that only around half of the 326 million Indian women living in poverty received some form of support under this fiscal transfer because of inadequate coverage (Pande et al. 2020). It was also found that those who got cash transfers had trouble accessing their money. These challenges perhaps led to a pivot in government strategy towards the use of the Aadhaar ecosystem to transfer cash. In May, there were already media reports that the government turned to other older datasets and schemes to channel fiscal relief, as it had become evident that there were challenges in identifying, targeting and reaching beneficiaries during COVID (Bhattacharjee 2020). A majority of migrant labourers, for instance, did not have the requisite documentation or the means to receive benefits through Aadhaar.

We attempt to understand the reasons for the failure of the digital panacea that Aadhaar and its associated programmes were touted to be, at the time when they were most needed – in the midst of what is likely to remain top among modern India's humanitarian tragedies. We begin with deconstructing the Aadhaar Enabled Payments System (AEPS), which is the financial bridge that is used to process social welfare transfers.

The mechanics of AEPS

AEPS builds on the unique ID framework to enable delivery of welfare transfers through an authentication process. AEPS is used to deliver cash at the last mile through Business Correspondents – agents of participant banks who operate micro-ATMs to authenticate and conduct transactions on behalf of end users or beneficiaries of government schemes. These Correspondents therefore substitute actual bank branches and ATMs. This Business Correspondent-based model was first envisioned in 2006, to plug the large gap in last-mile banking infrastructure through retail agents of banks in villages and towns. Initially, such correspondents were ex-government employees like postmasters, teachers and so on. Eventually the regulatory framework for such agents was further liberalised to include owners of kirana shops, ex bank employees and so on.

Under AEPS, a typical transaction involves input of an Aadhaar number at a micro-ATM operated by such a correspondent. Then the digitally signed and encrypted data packets are transferred via banks to the Unique ID Authority of India which is the home for the Aadhaar database. Once the Authority authenticates the user/beneficiary, by checking if the biometrics match, the concerned bank executes the requested transaction. To be clear however, there is physical handing over of cash from the Banking Correspondents who maintain a cash drawer, to the ultimate beneficiaries.

The AEPS model, with its reliance on a disaggregated network of retail agents and the Aadhaar authentication process has many benefits. These include the fact that most transactions are now agnostic of time and place, both important variables in the traditional brick-and-mortar banking world. For instance, close to 60 percent of AEPS transactions take place either during non-banking hours or on bank holidays, of which there are plenty (close to 40 every year in addition to weekends) (Balasubramanian et al. 2019).[1] It is also unsurprising that the most active AEPS users are those that inhabit some of the least banked regions and locales. A lack of adequate banking infrastructure in such areas, therefore, is seemingly no longer a constraint.

However, despite all its benefits, AEPS has a long history of trans-action failures, best documented by the Digital Identify Research Initiative at the Indian School of Business (ISB), which analysed a da-taset of around seven million AEPS transactions between 2014 and 2018. The research team classifies the types of common AEPS trans-action failures into those related to biometrics, as well as technical, and non-technical failures. Of these, they find an alarming overall failure rate of 34.03 percent – that is over a third of all AEPS transactions fail on account of one of the three underlying factors. And from within this, biometrics failures top the charts, accounting for 17.03 percent of fail-ures, followed by non-technical failures at 13.3 percent and technical failures accounting for 3.7 percent (Balasubramanian et al. 2019). Half of all the failures are on account of biometrics – the unique selling point of a digital unique ID like Aadhaar. Their inclusion into an identity tool to reduce chances of fraud and duplications, then, attests to the paradox at hand. That is, the very strength of a digital framework is also its greatest weakness. The taxonomy of these different types of failures, as identified by the team at ISB, is provided in Table 5.1.

There are multiple facets of interest that relate to the high rate of transaction failure on account of biometrics. The most prominent of those, in our view, is that biometrics seem to be least effective for the poorest income groups. For instance, the researchers at ISB, and several other academics, have noted that the chances of successful AEPS authentication using biometrics reduces drastically immediately after heavy manual labour. This means that those most in need – like

Table 5.1 Taxonomy of AEPS Transaction Failures

Biometric Failure
Biometric data did not match
Technical Failures
Switch not available
Database error
Reversal time-out
Socket connection error
System down
Transaction time-out
Non-Technical Failure
Insufficient funds in account: 51
Please seed your Aadhaar and mobile number with your bank account without fail for availing uninterrupted services
Transaction amount exceeded limit
Daily amount limit exceeded. Try tomorrow
Invalid account. Error: 52
Incorrect Pin

India's millions of migrants who toil away at construction, mining and manufacturing sites across the country – are the least likely to be able to successfully avail the benefits of AEPS at times when they would likely need it most, after a hard day of work.

Moreover, several studies have found that transaction failures are also associated with a rapid decline in trust of technological tools. In fact, this is also true of banking itself, which represents a poor quality of service ecosystem, wherein most bankers have little incentive or capacity to cater to the most disadvantaged sections of society. This is a vicious cycle, that leads to underutilisation of banking as a channel of social welfare delivery. And this is best evidenced by the fact that despite the Guinness World Record and other accolades that India has received for banking the unbanked under the Jan Dhan Yojana, a fifth of the bank accounts opened under the scheme remain dormant (*The Hindu Businessline* 2020). Moreover, most of the dormant accounts happen to be in the poorest states in the country, including Uttar Pradesh and Bihar, which between them account for more people than Central Europe. In these ways, the AEPS framework seems to militate against the very notion of Antyodaya.

The high rates of failures and low rates of trust proved to be a major reason for the ineffectiveness of AEPS, and concomitantly the Aadhaar-based social welfare ecosystem, at the peak of COVID. We subsequently examine these characteristic failures and how they specifically relate to the shortfalls in the emergency health response under COVID.

Transacting for survival

The demand for cash and welfare benefits surged in the immediate aftermath of COVID-linked lockdowns across India. About 403 million approved AEPS transactions were reported by the National Payments Corporation of India (NPCI), which operates the AEPS infrastructure in concert with the Unique ID Authority of India, and domestic banks, as compared to around 172 million transaction in the previous month (Raghavan and Shah 2020). According to analysts, this was due to a combination of drivers including the cash transfers under government schemes such as the Jan Dhan Yojana to vulnerable sections of society, including migrants who could no longer access bank branches and ATMs, even as they left cities in large numbers to go back to their towns and villages for sustenance. Business correspondents and AEPS became providers of last resort.

Unfortunately, increased demands on the AEPS infrastructure in April 2020 were accompanied with a large spike in transaction failure rates. This is perhaps best evinced by the fact that as compared to the ISB data which indicated an overall transaction failure rate of around 34 percent in pre-COVID times, AEPS was failing 39 percent of the time on average in April 2020 (Raghavan and Shah 2020). Some AEPS providers even reported that their transaction failure rates went up to 62 percent on the higher end of this spectrum. These stark failure rates beg the question – why was the Aadhaar-based financial ecosystem not cut out for the task when it was most needed? Was this some inherent failure of digital? What happened to all the arguments in favour of efficiency and inclusion, as they relate to Aadhaar and AEPS, that were recounted in the previous sections?

First, let's consider the unintended outcomes and paradoxes that are linked to the overall efficiency arguments, in the COVID emergency context. A key lesson, evident even in pre-COVID times, was that singular focus on reduction of identity fraud through a unique ID like Aadhaar, led to lack of focus on how to deal with another more entrenched form of fraud, known as quantity fraud. The scholar Reetika Khera defines this persistent form of leakage and corruption as "underselling" (Khera 2017). For instance, when Aadhaar is linked to the delivery of benefits under the Public Distribution Scheme (PDS) for food grains, it still does not solve for the fact that intermediaries such as the Business Correspondents at the last mile are meant to interact with the intended beneficiaries to authenticate and provide the benefits that the latter are entitled to. This means that PDS dealers that are similar to Business Correspondents for delivery of social welfare benefits, tend to provide less benefits than people are entitled to. For instance, they may provide 32 kgs of rice rather than 35 kgs to households, whereas they are meant to get the latter quantity. This scope for quantity fraud and corruption exists because the Aadhaar-based ecosystem does not account for the fact that beneficiaries have little agency to bargain when they depend on agents of the State to dole out their welfare benefits. Moreover, there are few accountability mechanisms or processes in place for such last-mile agents to act honestly. Similarly, Khera notes that another form of quantity fraud is well-documented in some of the poorer states like Bihar and Jharkhand, where such dealers also "siphon off entire months' worth of rations" and essentially skip certain months, since they are not really accountable to the beneficiaries and there is no effective grievance redressal process in place. It's not unimaginable to think that such quantity fraud can also take place when vulnerable migrants needed to access cash. In these ways, digital modes of welfare expose glaring gaps in state capacity and welfare scheme design, that tend to asymmetrically impact those at the bottom of the socio-economic pyramid. These are continuities rather than departures from the norm, and in many ways, and as illustrated throughout this book, thinking of digital as a silver bullet entrenches bad design.

The stark reality is that in the context of the COVID pandemic, during which millions of migrants were compelled to return home, lack of focus on quantity fraud took on a much larger dimension – it became a humanitarian issue. Moreover, reliance on biometrics as a means to reduce identity fraud in fact perpetuated and exacerbated such bad outcomes. For instance, Aadhaar enrolments were suspended in the initial days of COVID, precisely because of a dependence on biometrics (Kaur 2020). Similarly, there were a slew of transaction errors linked to mismatch of biometrics, that had no real solve if the intended beneficiary could not re-apply for Aadhaar. As mentioned earlier, manual labourers in particular are most prone to such errors because of changes in their fingerprints. And most of the transaction errors in April related to biometric mismatches, aside from timeouts of transactions because the digital backbone for AEPS was under considerable volume related stress (Raghavan and Shah 2020). Ideally such beneficiaries should have had the option of re-recording their fingerprints, with checks and balances in the AEPS framework to prevent this solution from inadvertently enlarging the scope for identity fraud. The fact that biometric failures were well-documented since the inception of Aadhaar and its associated financial and banking channels, should have spurred such basic design changes much earlier. For instance, these were recognised by the RBI's 2019 Committee on Deepening Digital Payments, chaired by one the key architects of Aadhaar, Nandan Nilekani. It was similarly also recognised in the Ministry of Finance's 2018 Report on Digital Payments, issued by the Ratan Watal Committee. But, as is par for the course in the developing world, such changes are rarely knowledge driven (they are usually only adverse event driven, if at all). At any rate, just the fact that the system is digitalised is not enough to prompt iterations based on stress tests. In fact, the digital characteristic perhaps fosters a converse impulse – to kick the redesign can down the road to a future date since there is so much political capital invested into its marketing as a perfect developmental solution.

Before we highlight some other challenges linked to the efficiency arguments in favour of AEPS, it may also be useful to point out another perverse consequence of a focus on biometrics for authentication. Because of the ubiquitous transaction failures that are now well-documented, and are linked to the quality of fingerprints and the underlying database, there is now a move towards facial recognition technology to overcome this challenge. Specifically, facial recognition is now seen as a "more advanced alternative, which is expected to reduce failure rates and take care of sanitisation issues pertaining to COVID" (Manikandan and Shukla 2020). This is despite the fact that facial recognition is a deeply contested technology, on the grounds that it invades individual privacy much more than any other form of authentication.

Another facet of the efficiency arguments in favour of digital is that digitalisation reduces the number of intermediaries in any welfare delivery ecosystem. However, this is far from the truth. For instance, Khera notes that the Aadhaar project has" spawned its own new army of middlepersons" including facilitators, seeding agents, data entry operators and so on. In fact, the ISB scholars correlate this to note that this creation of new intermediaries is problematic even for banks, let alone intended beneficiaries. Therefore, several banks now rely on aggregators to manage their business correspondents, creating an entirely new category of businesses or intermediaries. A key reason why this is the case is that digitalisation cannot gloss over the fact that most cash is still physical. This physicality is an important consideration because there is no acceptance infrastructure for digital cash, such as debit cards and payments wallets, across most villages and towns in India. Because the country falls woefully short on such infrastructure, analogous to banking infrastructure itself, digital is not a silver bullet. Consequently, even the RBI has noted that the percentage of cash withdrawals to GDP is at a constant in India at around 17 percent, despite digitalisation and a valiant attempt to reduce this through demonetisation of over 90 percent of currency notes in 2017 (RBI 2020).

To illustrate the continued centrality of cash in digital, it is useful to chart out the macro process flow under AEPS and a more traditional transfer of benefits to bank accounts of beneficiaries of social schemes. In the latter, money is simply transferred from the Consolidated Fund of India to the beneficiary's bank account, via the National Electronic Fund Transfer infrastructure managed by the RBI. Conversely, in the case of the AEPS, the National Payments Corporation of India (NPCI) also gets involved to act as the backbone of Aadhaar payments bridge and the AEPS. This new mode of welfare transfer, then, adds at least one more major intermediary to the mix, that is the NPCI, than the traditional mode. Moreover, as recounted earlier, AEPS does not eliminate the need for last-mile intermediation. Conversely, because of the inherent complexity of the biometrics-based digital ecosystem created around Aadhaar, there are several new physical intermediaries who have cropped up to support, and at times expropriate, the end user.

The proliferation of intermediaries in the digital scheme of things, highlights another unique gap that we have discussed in the case of agriculture in an earlier chapter. That is, intermediaries often lack capacity to deal with the intricacies of basic service delivery themselves, whereas the State tends to renege its responsibility to build such capacity because of the dazzle of digital. The negative impact of this gap became particularly prominent during COVID. That is, banking correspondents quickly ran out of cash in their registers/drawers, and they were unable to replenish due to physical restrictions on movement during the many central and state level lockdowns (Bhat et al. 2020). And there was no preferential access provided to such agents from rural areas by respective banks, simply because the situation was not accounted for by the State. Banks rarely have an incentive to go above and beyond their core business of lending, much less try to create guardrails for a state-mandated banking channel such as the Aadhaar-based financial services ecosystem. And the State on its part, shows a great deal of inertia to design for all possible failures of digital. In fact, just the opposite was noted during COVID by some of the Banking Correspondent aggregators. They reported harassment of their correspondents by the police, despite categorisation as "essential services", which should have allowed the agents to move about freely for replenishing their stocks of cash or carry out any other activity in their line of duty (Bhalla 2020).

The other plank for the glorification of digital is the purported benefit of greater inclusion or coverage. While we have already detailed several arguments that contest this from the perspective of efficiency of the AEPS, there are a few more issues to highlight that relate specifically to this pillar. As noted earlier, scholars from the Yale Economic Growth Centre highlighted that less than half of poor adult women were covered under the Jan Dhan Yojana. As a result, others suggested the use of alternative databases, such as the database of workers covered under the National Rural Employment Guarantee Scheme, to deliver fiscal relief during COVID. In fact, it seems even the State recognised the limitations of the Aadhaar infrastructure for targeting of beneficiaries, because by May 2020 it had started using the NREGS to transfer cash to rural beneficiaries instead (Bhattacharjee 2020). This was at best an implicit recognition, and not a public acknowledgement. Consequently, COVID has not engendered a much needed rethink on exclusionary nature of digital, especially when there is not enough focus on structural impediments to better service delivery.

We keep finding that the linkage of political signals to digital schemes is a major reason why there is not enough focus on good design of the latter. For instance, when the Jan Dhan Yojana for banking the unbanked was first launched in 2014, banks were given very ambitious targets to synchronise with the level of marketing effort put behind the scheme. This led to a frantic effort to meet targets, wherein several accounts were reportedly opened "without informed consent, duplicate accounts flourished, Aadhaar numbers were seeded without any safeguards, and so on" (Drèze and Khera 2020). This is inevitably linked to the dormancy of several of the accounts (up to 40 percent at a point in time before demonetisation), that led to the subsequent failure in targeting the most vulnerable citizens with benefits transfers using the Aadhaar database during the peak of COVID. The dazzle of digital unfailingly generates such paradoxes – the greater the political commitment to digitalisation, the less focus on challenges linked to accountability, exclusion and so on.

Overcoming the digital divide

To be clear, digital finance has the potential to increase efficiency and inclusion of social welfare schemes. But, this requires recognition of the fact that digital is no panacea for various limitations that are inherent to basic service delivery. These include low levels of financial literacy, limited or differential levels of access to digital infrastructure and devices, and the need to build capacity throughout delivery supply chains, including of intermediaries required for successful last-mile delivery. Moreover, it is also important to tabulate and assess the reasons for systemic failures in digital service delivery ecosystems, such as the level of service disruptions observed during COVID. While we have recounted some of these above, there are other aspects that are equally fundamental, such as the need for robust infrastructural backbones. India's poor quality of telecom infrastructure afflicts even the middle-class consumer, wherein transaction failures owing to network downtime or congestion are frequent. And finally, there is a universal need to integrate digital design thinking with economic incentives and related behavioural aspects of supply chain participants. For instance, social scheme beneficiaries must be empowered to hold their service providers accountable. Similarly, they require greater choice of providers, to switch seamlessly for a better quality of service. This applies to all tiers of basic service delivery, from the choice of banking partners, to network infrastructure, and ultimately, the last-mile intermediaries. However, greater competition for basic service delivery also requires that providers and agents are adequately compensated to reach the poorest of the poor, in the most remote and excluded regions of the country. This means that digital social welfare schemes are not ne-cessarily cheaper alternatives. If this is not realised and designed for, technology can end up exacerbating the divide between the haves and the have nots, a common symptom of digitalisation.

India is not the only country in the world that banks on digital as a means to lessen the gap between haves and have nots. For instance, Kenya, in its design of its government to citizen financial services scheme, accounts for key aspects such as "allowing recipients to choose their payment delivery providers, empowering recipients to expect better customer service as banks competed for their business" and so on (McKay and Gcinisizwe 2020). Similarly, it compensates such providers with a tiered scale, "offering the highest fees for the most remote areas". These are fundamental design fixes that India may consider adopting in its own heterogenous and challenging social welfare and digitalisation context. The importance of these measures became prominent during COVID, but all changes to design need not be led by adverse events. India was not alone in this learning. For instance, even in smaller countries with less complex delivery mechanisms, such as Ecuador, a key learning of COVID was that there was a large amount of "crowding in at cash-out points" (Gentilini et al. 2020). In response, several countries moved to account-based solutions, where cash transfers were made into accounts in the name of specific recipients and that were maintained by payment service providers. Of course, this also creates large scope for both identity and quantity fraud, which must be mitigated at the blue-print stage itself, through appropriate accountability measures.

One accountability measure that has universal appeal and is adopted in various jurisdictions is the standardisation of consumer grievance redressal frameworks or processes linked to financial services more broadly, and social welfare schemes more specifically (Gupta 2021). Normally, policymakers tend to assume that the existing scaffolding of supervisory institutions and enforcement agencies, is enough to deal with such exigencies. In fact, we find throughout the course of this book that accountability is at best an afterthought. However, in the developing country context, most agencies ordinarily responsible for such oversight are already stretched for capacity, and are not equipped to deal with the combination of higher volumes, greater complexity, and faster evolution of digital service delivery ecosystems. Aside from more attention to de-sign for digital, this merits greater focus on tiered, transparent and dedicated systems to channel grievances to regulatory bodies such as the RBI. And such systems need to be supplemented with auditable public registers or data repositories that are open to scrutiny so that the process of design and iteration is not bereft of external input. Without such citizen-centricity, it is unlikely that any digital solution can pass muster in a future emergency-like situation as the one we saw under COVID. And as we have maintained here, this means that digital systems will continue to fail when they are most needed, and fail those that most need them. This will consequently erode hard-to-earn trust in technology, particularly at the bottom of the socio-economic pyramid. It is unlikely that such citizens as migrant labourers, will repose their trust in the AEPS if they have borne the brunt of its failures during COVID.

Note

1 Close to 40 every year in addition to weekends.

Bibliography

Akinbi, Alex, Mark Forshaw, and Victoria Blinkhorn. "Contact Tracing Apps for the COVID-19 Pandemic: A Systematic Literature Review of Challenges and Future Directions for Neo-Liberal Societies." *Health Information Science and Systems* (2021). 10.1007/s13755-021-00147-7

"Almost Every Fifth Jan Dhan Account 'inoperative'." *The Hindu BusinessLine*, February 06, 2020. https://www.thehindubusinessline.com/money-and-banking/almost-every-fifth-jan-dhan-account-inoperative/article30754738.ece (accessed on April 25, 2021).

"Assessment of The Progress of Digitisation From Cash To Electronic." Reserve Bank of India, February 24, 2020. https://www.rbi.org.in/Scripts/PublicationsView.aspx?id=19417

Balasubramanian, Padmanabhan et al. "Fintech For The Poor: Do Technological Failures Deter Financial Inclusion?." Indian School of Business, Digital Identity Research Initiative, April 01, 2019. https://diri.isb.edu/en/community/blog-grid/alternatives-to-aadhaar-based-biometrics-in-the-public-distribut1.html

"Bank Accounts Opened Under Pradhan Mantri Jan Dhan Yojana Crosses 40-Crore Mark." *Economic Times*, August 03, 2020. https://economictimes.indiatimes.com/industry/banking/finance/banking/bank-accounts-opened-under-pradhan-mantri-jan-dhan-yojana-crosses-40-crore-mark/articleshow/77328277.cms?from=mdr (accessed on April 25, 2021).

Bassi, Abhinav et al. "An Overview Of Mobile Applications (Apps) To Support The Coronavirus Disease 2019 Response In India." *Indian Journal of Medical Research* 151 (2020). https://www.ijmr.org.in/article.asp?issn=0971-5916;year=2020;volume=151;issue=5;spage=468;epage=473;aulast=Bassi

Bhalla, Tarush. "Coronavirus: Rural India to Face Cash Shortage Amidst Lockdown, Says BCFI." Yourstory, March 27, 2020. https://yourstory.com/2020/03/rural-india-cash-shortage-bcfi-coronavirus

Bhat, Sunil et al. "Addressing Three Key Issues Of BC Agents In India For COVID-Like Challenges." MicroSave Consulting, Blog, 2021. https://www.microsave.net/wp-content/uploads/2021/01/210105_Blog_Addressing-three-key-issues-of-BC-agents-in-India-for-COVID- ... .pdf (accessed on April 25, 2021).

Bhattacharjee, Subhomoy. "Fiscal Stimulus 2.0: Why Cash Support To Migrant Workers Was Aborted." *Business Standard*, May 18, 2020. https://www.business-standard.com/article/economy-policy/fiscal-stimulus-2-0-why-cash-support-to-migrant-workers-was-aborted-120051800047_1.html (accessed on April 25, 2021).

"Communicable Disease Surveillance And Response Systems." World Health Organisation, 2006. https://www.who.int/csr/resources/publications/surveillance/WHO_CDS_EPR_LYO_2006_2.pdf?ua=1

Colesca, Sophia. "Understanding Trust in E-Government." *Inzinerine Ekonomika* 63, no. 3 (2009). https://www.inzeko.ktu.lt/index.php/EE/article/view/11637

Deb, Sidharth. "Privacy Prescriptions For Technology Interventions On COVID-19 In India." Internet Freedom Foundation, Working Paper No. 3, 2020. https://docs.google.com/document/d/1nDoPzygQyTetEguOlzula5O9y5f3f5YJDsA2Pd99O6U/edit#heading=h.5je3mbv90ww7

Drèze, John, and Reetika Khera. "Getting Cash Transfers Out Of A JAM." *The Hindu*, May 13, 2020. https://www.thehindu.com/opinion/lead/getting-cash-transfers-out-of-a-jam/article31568674.ece (accessed on April 25, 2021).

Dutt, Anonna. "20% Beds In 117 Private Hospitals To Be Reserved For COVID-19 Surge." *Hindustan Times*, May 25, 2020. https://www.hindustantimes.com/india-news/20-beds-in-delhi-s-117-pvt-hospitals-to-be-reserved-for-covid-patients/story-IIOpUI6p1wXFgAzuRIXnbK.html (accessed on April 25, 2021).

Gale, Jason. "61-Year-Old Patient Is First to Die in Wuhan Pneumonia Outbreak", *Bloomberg Quint*, January 14, 2020. https://www.bloombergquint.com/onweb/wuhan-doctors-tried-to-save-pneumonia-patient-with-life-support (accessed on April 25, 2021).

Gallagher, Gerard and Peter Hotez. "Zika Caused More Birth Defects in US than Expected." *Thorofare* 31(2): 4, 2018. https://www.proquest.com/openview/035b33ca5e8f209e7a45c19c53ee7257/1?pq-origsite=gscholar&cbl=29327

Garari, Kaniza. "ICMR Shocker: India Missed 80 Corona Cases for Every One Detected." *The Deccan Chronicle*, December 9, 2020. https://www.deccanchronicle.com/nation/current-affairs/120920/icmr-shocker-india-missed-80-corona-cases-for-every-one-detected.html

Gentilini, Ugo et al. "Social Protection and Jobs Responses to COVID-19: A Real-Time Review of Country Measures." Living Paper, Version 10, May 22, 2020. https://www.ugogentilini.net/wp-content/uploads/2020/05/Country-SP-COVID-responses_May22.pdf

"Global Spending on Health: A World in Transition." World Health Organisation. 2019. https://www.who.int/health_financing/documents/health-expenditure-report-2019.pdf?ua=1#:~:text=It%20was%20US%24%207.8%20trillion,US%24%207.6%20trillion%20in%202016.&text=The%20health%20sector%20continues%20to,economy%20grew%203.0%25%20a%20year

"Govt Pulled Up by RTI Body Over Evasive Reply On Aarogya Setu. It Clarifies." Hindustan Times, October 28, 2020. https://www.hindustantimes.com/india-news/govt-pulled-up-by-rti-body-over-evasive-reply-on-aarogya-setu-it-clarifies/story-C8K1A3tXrl7wCPvcv7fxSP.html (accessed on April 25, 2021).

Gupta, Aarushi. "Proposing a Framework to Document Exclusion in Direct Benefit Transfers." DVARA Research, February 11, 2021. https://www.dvara.com/blog/2021/02/11/proposing-a-framework-to-document-exclusion-in-direct-benefit-transfers/

Hindustan Times. "States Did Not Share Data on Oxygen Shortage Deaths during Covid: Centre." April 6, 2022. https://www.hindustantimes.com/india-news/states-did-not-share-data-on-oxygen-shortage-deaths-during-covid-centre-101649187149023.html

International Health Regulations. (World Health Organization) 2005. https://www.who.int/publications/i/item/9789241580410

Kaur, Harvinder. "Suspension Of Aadhaar-Related Services Leaves Scheme Beneficiaries In Ludhiana Helpless." *Hindustan Times*, May 29, 2020. https://www.hindustantimes.com/cities/suspension-of-aadhaar-related-services-leaves-scheme-beneficiaries-in-ludhiana-helpless/story-9EATmq4EOTYtnaBR7wwaYM.html (accessed on April 25, 2021).

Khera, Reetika."Impact of Aadhaar in Welfare Programmes", *Economic and Political Weekly* 52, no. 50 (2017): 61–70. https://www.researchgate.net/publication/322151210_Impact_of_Aadhaar_in_Welfare_Programmes

Kumar, Devesh. "Half a Million COVID-19 Cases in India: How We Got To Where We Are." The Wire, June 2020. https://thewire.in/covid-19-india-timeline (accessed on April 25, 2021).

Manikandan, Ashwin, and Saloni Shukla. "Facial Recognition, Iris Scans May Be Used For Welfare Scheme Payouts." *Economic Times*, August 26, 2020. https://economictimes.indiatimes.com/corporate/facial-recognition-iris-scans-may-be-used-for-welfare-scheme-payouts/articleshow/77753699.cms (accessed on April 25, 2021).

McKay, Claudia, and Gcinisizwe Mdluli. "Kenya's Expansion of G2P Becomes Lifeline During COVID-19 Crisis." CGAP, Blog, April 29, 2020. https://www.cgap.org/blog/kenyas-expansion-g2p-becomes-lifeline-during-covid-19-crisis (accessed on April 25, 2021).

"National Capacities Review Tool for a Novel Coronavirus." World Health Organisation, January 09, 2020. https://www.who.int/publications/i/item/national-capacities-review-tool-for-a-novelcoronavirus

OECD. "Health at a Glance 2021: OECD Indicators." *Health at a Glance OECD*, 2021. doi: 10.1787/ae3016b9-en.

"On The Record: Delhi Doctor Calls Govt's Moves 'Kafkaesque'." *Hindustan Times*. June 09, 2020. https://www.hindustantimes.com/videos/on-the-record/on-the-record-delhi-doctor-calls-govts-moves-kafkaesquevideo/video-T3BU5Xw3o0pBNxQfcUWe4J.html (accessed on April 25, 2021).

Pande, Rohini et al. "A Majority of India's Poor Women May Miss COVID-19 PMJDY Cash Transfers." Yale Economic Growth Centre, April 2020. https://egc.yale.edu/sites/default/files/COVID%20Brief.pdf

"PM's Reply to The Motion Of Thanks On The President's Address In Lok Sabha." Prime Minister's Office, Government of India, February 10, 2021. https://pib.gov.in/PressReleaseIframePage.aspx?PRID=1696848

"Pradhan Mantri Garib Kalyan Package: Progress so Far." *Press Information Bureau*, Government of India, April 23, 2020. https://pib.gov.in/PressReleseDetail.aspx?PRID=1617393

Raghavan, Malavika and Samir Shah. "Fix the Problems In Aadhaar-Based Cash Transactions." *The Mint*, May 08, 2020. https://www.livemint.com/opinion/columns/fix-the-problems-in-aadhaar-based-cash-transactions-11588930862806.html (accessed on April 25, 2021).

Rajagopal, Krishnadas. "Aadhaar Gets Thumbs Up from Supreme Court." The Hindu, September 26, 2018. https://www.thehindu.com/news/national/aadhaar-gets-thumbs-up-from-supreme-court/article25051538.ece (accessed on April 25, 2021).

Singh, Manmohan. Speech at the Civil Services Day. Delhi, April 21, 2013. https://archivepmo.nic.in/drmanmohansingh/speech-details.php?nodeid=1308

Singh, Karn Pratap. "Delhi Police to Check Aarogya Setu App of All Visitors to Police Buildings." *Hindustan Times*, October 24, 2020. https://www.hindustantimes.com/delhi-news/delhi-police-to-check-aarogya-setu-app-of-all-visitors-to-police-buildings/story-zbMScxPMB8TqA02OB8iwTK.html (accessed on April 25, 2021).

Surveillance Strategies For COVID-19 Human Infection." World Health Organisation, June 05, 2020. https://www.who.int/docs/default-source/coronaviruse/risk-comms-updates/update-29-surveillance-strategies-for-covid-19-human-infection.pdf?sfvrsn=3c2cab92_2

Verma, Shubham. "Aarogya Setu Now World's Most Downloaded Covid-19 Tracking App." *India Today*, July 16, 2020. https://www.indiatoday.in/technology/news/story/aarogya-setu-now-world-s-most-downloaded-covid-19-tracking-app-1701273-2020-07-16

Vyas, Mahesh. "Job Losses in White And Blue Collar Workers." Centre for Monitoring Indian Economy, *Economic Outlook*, September 14, 2020. https://www.cmie.com/kommon/bin/sr.php?kall=warticle&dt=2020-09-14%2021:47:53&msec=416

6 Conclusion – Stepping Back to Leap Forward

Digital technology is increasingly interacting with our physical reality – a characteristic that will only enhance with the proliferation of "cybernetic systems" that can learn and potentially act independently of human beings. Norbert Wiener laid the theoretical foundations of cybernetics with his influential book called *Cybernetics or Control and Communication in the Animal and the Machine* in 1948. Cybernetic systems are complex information systems that can compute information and even design their own code to reach from a starting point to an end point by learning and acting on a strategy. A rudimentary everyday example is the use of digital maps on a smartphone that present the user with a "current state" or a static location, as well as a representation of the "end state" which could be the desired end location, and a strategy to get there. A most familiar example, however, is the human brain – which is also a goal-directed information system. And like the human brain, in the best case, these amazing digital systems will help us solve society's deepest challenge, and in the worst case, they will exacerbate them.

Wiener's seminal work on information systems was as philosophical as it was technical, because it was able to foresee several failings of a modernity mediated via technology. We rediscover many such predicted failings in the Indian context. For instance, in 1950, Wiener said "the idol is the gadget and I know very great engineers who never think further than the construction of the gadget and never think of the question of integration between the gadget and human beings in society".[1] This sounds remarkably similar to the ethos of India's governing elite, who are trapped in echo chambers of techno-evangelism, and seek to digitise our societies and computerise our governance, without adequate reflection on the purpose of such transformation.

We have attempted to demonstrate the unintended consequences of digital solutionism and determinism based on the current generation of

DOI: 10.4324/9780429324901-6

digital technology, in both the public and private sectors. Since digital technology is rapidly advancing, we soon expect there to be greater disintermediation of people and institutions from the functioning of such technology, as Norbert Wiener had predicted. We therefore stand at a critical crossroad as a society. We must identify our blind spots and remedy our impulse to treat digital technology as a cure all, before it become ubiquitous and adds so many layers of complexity to existing challenges that we lose perspective permanently.

Most of our governing elite take a maximalist view of digital technology. That is, they tend to see more as better; whereas, increasingly complex development and growth challenges require a more calibrated approach. What makes things worse is that digital technologies tend to homogenise individuals, firms and markets; much like what politicians and administrators tend to do. In the case of digital technology, this tendency to homogenise stems from the engineering of software and code to identify and group similar things. However, such maximalism and homogenisation militate against equitable progress because it is exclusionary by design. It ignores people and producers on the margins. It condemns those left behind permanently. It reinforces existing binaries of access. It inverts the logic of Antyodaya.

Perhaps, the only remedy is to take a step back and reflect on means to design and guide the interaction of such technology with society and markets, to achieve sensible goals. This is easier said than done, but reliance on first-principles may help.

Design for development, not for digital

The poor design choices that accompany the rollout of digital technology, often result from a lack of context-specificity. This is exemplified by the Mkisan programme, a text-based messaging service for farmers, a majority of whom are illiterate. That Mkisan is only available in English underscores that the circumstances peculiar to the individual were not considered. Mkisan excludes all, but a very small minority of the population knows how to read and speak in English. Similarly, the use of biometrics as a means of authentication for the delivery of social welfare benefits doesn't reflect adequate sensitivity to the context of beneficiary communities. We have discussed how biometric authentication is susceptible to failure in the case of those engaged in manual labour – individuals who needed the most support as seen during the COVID crisis. But there is no impetus towards iterative design of such authentication frameworks, because of the false assumption that a majority of leakages in welfare subsidies are due to identity fraud rather than quantity fraud. However, beneficiaries often get less than they are owed because the Aadhaar ecosystem does not guard against corruption at the last mile. Digital technology cannot overcome dependence on intermediaries, unless entire systems of public service delivery are rebuilt to be auditable, accountable and corruption free.

A lack of context-appropriate policy design is also visible in popular discourses about the market potential of data. Many in the private and public sectors feel there is causal relation between the abundance of data and economic progress. However, we find that this view over-looks the important journey of more advanced economies and firms. The countries and companies that derive the most value from data, invest in its processing and not just its collection. More importantly, several of the gains afforded by data stem from a robust industrial edifice that often lies beneath digital scaffolding. Digital technology optimises operations in the traditional commercial ecosystem, and further innovation is often built on the back of this. What follows is that the gains from the digital economy are likely to be minimal if the undergird of the real economy is weak.

Therefore, we suggest a three-pronged, principles-led design framework to optimise our use of digital technology.

The first pillar is empowerment-led design. Empowerment-led design entails a shift in focus from access to technology, to enhanced enablement of the individual, to achieve desired outcomes. The paradoxes thrown up because of the dazzle of digital, result from the inability of technology-driven development schemes to provide intended beneficiaries with true agency. They tend to follow formulaic recipes for success, such as maximisation of the number of beneficiaries reached, and conversely undervalue outcomes that may are not easily quantified such as the number of beneficiaries enabled. Empowerment-led design, on the other hand, prompts decision makers to understand the complex perspective of a beneficiary and assess the promise and peril of an intervention from her perspective.

Empowerment-led design enables three important outcomes. One, it gives individuals greater agency since interventions would require customisation to serve specific needs. A good example here are Farmers Field Schools, a form of extension devised by the Food and Agricultural Organisation that is geared towards teaching "farmers how to experiment and problem-solve independently, through interactive learning and field experimentation, with the expectation that they will thus require fewer extension services and will be able to adapt the technologies to their own specific environmental and cultural needs" (Ramanjaneyulu et al. 2009). Farmers Field Schools have enjoyed great success in promoting ecological practices such as integrated pest management, that allow farmers to manage pests sustainably without compromising on yield. Integrated pest management is based on the premise that pests are a "symptom of an ecological disturbance rather than the cause" of it in farming (Ramanjaneyulu et al. 2009). Integrated pest management through Farmers Field Schools or IPM-FFS as it is known has enjoyed significant success in India such as Telangana where it saved a majority of farmers from re-sowing their crops due to an infestation of the Red Hairy Caterpillar.

According to the Tamil Nadu Agricultural University, a key constraint of FFS is the limitation of capacity to scale these projects. Digital technology could play a role here, provided its interactivity is adequately harnessed and its introduction is supplemented by the enhancement of capacities of trainers and farmers through digital literacy programmes.

Two, empowerment-led design also ensures that it vests enough bargaining chips in the hands of beneficiaries to ensure they can demand what is owed to them. Essentially, it creates a system of checks and balances that enables them to demand service delivery from those responsible for meting it out to them. The limitations on connectivity at the last mile, namely those of institutional capacity, digital literacy and quality of connectivity infrastructure, means that intermediaries will remain an inexorable component of public service delivery. The State could tip the balance of power in favour of the beneficiary or recipient by creating structures of downward accountability.

A system of downward accountability rests on upending the mechanisms of information symmetry, implementing accessible grievance redressal and inclusive participation. For instance, for the implementation of the National Rural Employment Guarantee Scheme (NREGA) in the state of Andhra Pradesh, two steps were undertaken to empower workers. The Andhra Pradesh Government computerised all information related to the workings of the scheme and made it publicly available, to counter abuses related to the transactions involved in NREGA, namely the payment of wages and the hiring of workers (Masiero and Maiorano 2016). Further, the political class promoted the "formation of Shrama Shakti Sanghas (fixed labour groups) at the village level" (Dutta 2015). Shrama Shakti Sanghas are small groups of around 20 NREGA wage seekers and panchayat members who train poor NREGA labourers and help them unionise. The labourers, in turn, pay the group a small fee for protecting their rights. The work done by Shrama Shakti Sanghas enhances "workers' awareness and political sensibilities, and helps them deal with the administration to express their grievances as a group" (Dutta 2015). Though the combination of the Shrama Shakti Sanghas and open NREGA data are steps in the right direction towards key gaps in the interplay of these initiatives prevent them from empowering workers and providing downward accountability (Masiero et al. 2015). Computerisation of NREGA payments is only until the banks and post-offices that disburse the payments. Thus, the most crucial phase of NREGA, namely the process of making the payments is not transparent. Concomitantly, though the Sanghas had some success in granting agency to workers, their structure tends to be hierarchical and power concentrates in the hands of a few within the groups. Efforts may be considered, then, towards computerising the process of payment disbursal and establishing ways of getting that information in the hands of the labourers themselves and bringing in grievance redressal mechanisms to enable workers to call out abuses in both the Shrama Shakti Sanghas and the payment disbursal agencies.

Three, empowerment-led design breaks the substantialist tendencies that ignore problems that are not neatly quantifiable. Under empowerment-led design, recipient welfare is the core metric, which requires to be gauged through both qualitative and quantitative lens.

The second pillar is encouragement-based design. Encouragement-based design works on a system of subtle incentives and nudges to achieve desired outcomes. It can work in tandem with empowerment-based design, through focus on creating and aligning incentives for different stakeholders within an ecosystem, to ensure everyone benefits by doing what they are meant to do. Encouragement-based design is an important consideration in a country like India, where intermediaries form an inexorable component of last-mile public service delivery. Our reliance on intermediaries endures due to gaps in state capacity as well as physical infrastructure such as agricultural markets, last-mile bank branches and good quality telecom networks. Therefore, encouragement-based design can help stakeholders devise safeguards and incentives to ensure that delivery chains that involve intermediaries remain corruption free.

The private sector tends to rely on encouragement-based design, and therefore there are valuable lessons to be learnt from it. Take, for instance, the way a private credit card network works. Credit card providers tie up with banks and merchants to deliver seamless transaction services to consumers. These networks rely on incentives at each node of the supply chain. Banks share the transaction fee, merchants get access to a larger customer base and consumers or users don't need to withdraw and handle cash. Similarly, encouragement-based design may also help the State to incentivise individuals to sign up more freely for its schemes. Encouragement-based design necessitates better communications of the welfare benefits associated with development schemes. Illustratively, Kerala instituted a scheme for migrant workers that tethers welfare benefits that include the payment of INR 2 lakh as compensation to the worker's family upon his death, to the all-important act of registration.

The third pillar is participatory design, which can help overcome the common entrapment of decision makers by the dazzle of digital technology. Development initiatives in the physical realm are required to consider implications related to people on the ground. Undertaking to build a road, for instance, necessitates that the State consider the interests of citizens in myriad ways – their movements, where they need to get to, what kind of vehicles will they move around in, the timing of their travel, whether their settlements would be affected by the construction etc. There is thus a sense of collective responsibility because everyone is a stakeholder in the making of the road. As a result, citizens have some claim to its ownership because it affects their day-to-day life.

Digital networks, on the other hand, disintermediate people, as they are built to transmit bytes of data across terminals. Digital technology is brought in to remove people from processes to make distribution or production more efficient, and is often packaged and sold to the political class by technocrats who are far removed from the ground. As such, it is much easier for the State to claim end-to-end ownership over digital interventions. The consequent classification of digital interventions as a proprietary or exclusive concern of the State or the political class is problematic for several reasons. Decision makers tend to take critiques of their digital solutions personally, because of a concentrated and non-inclusive sense of ownership. Such characterisation is innate and reduces the scope for course correction. The personalisation of the digital also goads a tendency for the State to establish eminent domain and erode private property rights in the digital realm. Thus, as we saw in Chapter 4, the State introduces protectionist policies to ringfence what it considers as its sovereign turf. Additionally, such hyphenation prompts a lack of regard for any notions of citizen empowerment and encouragement as well as the many behavioural factors that determine the success of digital schemes, such as wide mistrust of new technology.

Participatory design of schemes that rely on digital technology can help calibrate the latter to achieve national goals, economic, social and cultural. It necessitates citizen participation and establishes the basis for a more inclusive digital ecosystem, not plagued by the same inequities as the physical realm. Participatory design combines democratic values such as trust and transparency with the practical aspects of democratic processes such as citizen feedback and consensus. It bodes for better outcomes in the long run because the panoply of participants is allowed to dissect technology and question its benefits. Most importantly, participatory design fortifies the pillars of empowerment and encouragement-based design.

Finally, we must caution that even the triad of empowerment, encouragement and participation-based design is necessary but not sufficient to ensure that the use of digital technology optimises welfare. The only way technology will live up to its promise is through a broad-based recognition that its use necessitates greater investments in state capacity, not less. It requires value accretive competition in the private sector, not protectionism that stifles innovation. And it demands vigorous civil society engagement, and is not an impulse to simply to seal the social contract with digital technology's emancipatory promise. All of this, in turn, requires an enlightened leadership at the top rungs of politics, government, industry and civil society. One that recognises the need to step back and think deeply about the end goals and purposes of digital technology. Such reflection will allow the country leap forward to a modernity that is not inconsistent with the progress and prosperity of all citizens, irrespective of their current state.

Note

1 https://www.wnyc.org/story/norbert-wiener-conscientous-gadgeteer/

Bibliography

"1,000 More Mandis to be Integrated with e-NAM in 2021–22: Govt." *Times of India*, February 04, 2021 http://timesofindia.indiatimes.com/articleshow/80694357.cms?utm_source=contentofinterest&utm_medium=text&utm_campaign=cppst (accessed on April 25, 2021).

Dutta, Sujoy. "An Uneven Path To Accountability: A Comparative Study Of MGNREGA In Two States Of India." WZB Discussion Paper, No. SP I 2015-201,Wissenschaftszentrum Berlin für Sozialforschung (WZB), Berlin, 2015. https://www.econstor.eu/bitstream/10419/108736/1/820319341.pdf

Masiero, Silvia, and Diego Maiorano. "Empowering Wageseekers? The Computerisation of India's NREGA in Andhra Pradesh." Blog, London School of Economics, 2016. http://eprints.lse.ac.uk/74693/1/blogs.lse.ac.uk-Empowering%20wageseekers%20The%20computerisation%20of%20Indias%20NREGA%20in%20Andhra%20Pradesh.pdf (accessed on April 25, 2021).

Ramanjaneyulu G., M. Chari, T.V. Raghunath, Z. Hussain, and K. Kuruganti. "Non Pesticidal Management: Learning from Experiences." In: Peshin R., Dhawan A.K. (eds) *Integrated Pest Management: Innovation-Development Process*. Dordrecht: Springer, 2009. 10.1007/978-1-4020-8992-3_18

Index

For Product Safety Concerns and Information please contact our EU
representative GPSR@taylorandfrancis.com
Taylor & Francis Verlag GmbH, Kaufingerstraße 24, 80331 München, Germany